Advanced and Intelligent Manufacturing in China

Series Editor

Jie Chen, Tongji University, Shanghai, Shanghai, China

This is a set of high-level and original academic monographs. This series focuses on the two fields of intelligent manufacturing and equipment, control and information technology, covering a range of core technologies such as Internet of Things, big data, 3D printing, robotics, intelligent equipment, industrial network security, and artificial intelligence, and epitomizing the achievements of technological development in China's manufacturing sector.

With Prof. Jie Chen, a member of the Chinese Academy of Engineering and a control engineering expert in China, as the Editorial in Chief, this series is organized and written by more than 80 young experts and scholars from more than 40 universities and institutes. It typically embodies the technological development achievements of China's manufacturing industry. It will promote the research and development and innovation of advanced intelligent manufacturing technologies, and promote the technological transformation and upgrading of the equipment manufacturing industry.

Jinfeng Wang

Intelligent Manufacturing System and Intelligent Workshop

Parameter Optimization, Process Planning, Workshop Scheduling, Typical Case

Chemical Industry Press Co., Ltd.

Springer

Jinfeng Wang
Department of Mechanical Engineering
North China Electric Power University
Baoding, Hebei, China

ISSN 2731-5983 ISSN 2731-5991 (electronic)
Advanced and Intelligent Manufacturing in China
ISBN 978-981-99-2010-5 ISBN 978-981-99-2011-2 (eBook)
https://doi.org/10.1007/978-981-99-2011-2

Jointly published with Chemical Industry Press
The print edition is not for sale in China (Mainland). Customers from China (Mainland) please order the print book from: Chemical Industry Press.

This Springer imprint is published by the registered company Springer Nature Singapore Pte Ltd.
The registered company address is: 152 Beach Road, #21-01/04 Gateway East, Singapore 189721, Singapore

Contents

Chapter 1
Introduction to Intelligent Manufacturing

1.1 Overview of Intelligent Manufacturing

Intelligent Manufacturing (IM) is derived from the research achievements of Artificial Intelligence. Intelligent Manufacturing system is a human–machine integrated intelligent system composed of intelligent machines and human experts. The system can carry out intelligent activities such as analysis, reasoning, judgment, conception, and decision-making in the manufacturing process. It also can expand, extend and partially replace the mental work of human experts based on the cooperation between people and intelligent machines. Intelligent manufacturing has updated the concept of automated manufacturing, expending it towards flexibility, intelligence and high integration.

The "'Twelfth Five-Year' Special Plan for the Intelligent Manufacturing Technology Development" released in March 2012 points out that Intelligent manufacturing technology is based on modern sensing technology, network technology, automation technology, anthropomorphic intelligent Technology and other advanced technologies. Through intelligent perception, human–computer interaction, decision-making and execution technology, the intelligence of design process, manufacturing process and manufacturing equipment is achieved, which reflects the deep fusion and integration of information technology, intelligent technology and equipment manufacturing technology. Therefore, intelligent manufacturing technology refers to an advanced manufacturing technology. It uses computers to simulate the analysis, judgment, reasoning, conception and decision-making of manufacturing experts. The integration of the above intelligent activities and intelligent machines is applied throughout various subsystems of manufacturing enterprises. This technology enables the high degree of flexibility and integration of the entire manufacturing enterprise operation, replaces or extends part of the mental work of experts in the manufacturing environment, and collects, stores, improves, shares, inherits and develops the experienced information of manufacturing experts. The intelligent manufacturing technology greatly improves production efficiency.

J. Wang, *Intelligent Manufacturing System and Intelligent Workshop*,
Advanced and Intelligent Manufacturing in China,
https://doi.org/10.1007/978-981-99-2011-2_1

Intelligent manufacturing system is also a manufacturing system that consists of intelligent manufacturing units with partly or fully autonomy and cooperation, and that exhibits human-like intelligent behavior throughout manufacturing activities. It is the integration and fusion of advanced manufacturing technology, information technology and intelligent technology in equipment products, reflecting the intelligence, digitization and networking of the manufacturing industry. Its main feature lies in the acquisition, expression and use of knowledge in the work process. According to its knowledge sources, intelligent manufacturing systems can be divided into two categories:

- Non-autonomous manufacturing systems represented by expert systems. The knowledge of this type of system is summed up from the empirical knowledge of human manufacturing.
- Autonomous manufacturing system based on self-learning, self-evolution and self-organization of the system. This kind of system can continuously learn, improve and evolve the original knowledge independently in the working process, and has strong adaptability and highly open innovation ability. With the development of artificial intelligence technology represented by deep learning technologies such as neural networks and genetic algorithms, intelligent manufacturing systems are gradually transitioning from non-autonomous intelligent manufacturing systems to autonomous systems with self-learning, self-evolution and self-organization.

1.2 Core Technologies of Intelligent Manufacturing

Intelligent Manufacturing includes the following core technologies: industrial robots, 3D printing, radio frequency identification (RFID), wireless sensor network (WSN), Internet of Things (IoT) and cyber physical system (CPS), industrial big data, cloud computing, virtual reality (VR), and artificial intelligence (AI). Among the core technologies, industrial robots and 3D printing are the two major hardware tools, RFID and WSN are the two main means of communication used for interconnection, IoT, Industrial Big Data and Cloud Computing are the three main foundations based on distributed analysis and decision-making, while VR and AI are the two future-oriented traction technologies.

1.2.1 Industrial Robot

International Federation of Robotics (IFR) defines a robot as follows: a robot is a machine that works semi-autonomous or fully autonomous to work that is beneficial to human beings. It is called the industrial robot when applied to the production process, called a special-purpose robot (special robot) in special environments, and called (housekeeping) service robots in the family or directly serving person [1].

The International Organization for Standardization (ISO) defines a robot as "a robot is an automatic, position-controllable, programmable, multifunctional manipulator with several axes capable of handling various materials, parts, tools, and specialized devices to perform various tasks with the aid of programmable operations" [2]. According to the ISO definition, an industrial robot is a multi-degree-of-freedom robot oriented to the industrial field, a machine device that automatically performs various functions by its own power and control capabilities [3]. It accepts human instructions and operates following the set program. Typical applications of industrial robots include welding, painting, assembly, collection and placement (such as packaging and palletizing, etc.), product inspection and testing, etc. [4]. According to the development roadmap of robots released by the United States in March 2013 [5], mobile and operable equipment and devices with certain intelligence are called robots. The intelligent hoovers, intelligent lawn mowers, intelligent homes and Google unmanned vehicles, are considered robots. Industrial robots are often divided into the following categories [6, 7]:

(1) **Mobile robot (AGV)**

Mobile robot (AGV) with the function of mobile, automatic navigation, multi-sensor control is widely used in flexible handling, transmission and other functions in machinery and other industries, and is also used in automated three-dimensional warehouses, flexible processing and assembly systems. It can be used as a transport tool in the sorting of items in airports and stations, which is an important part of intelligent internet of things and intelligent warehouses.

(2) **Welding robot**

On the assembly line of automobile production, welding robots are used to replace humans in automatic welding work, which includes spot welding robots and arc welding robots. Spot welding robots are mainly used for welding works of whole automobiles, 165 kg spot welding robots are the most commonly used type of robots in the current automobile welding. Arc welding robots are mainly used in the welding production of various auto parts. Welding robots have the characteristics of stable performance, large working space, fast movement speed and strong load capacity, etc.. Its welding quality is obviously better than that of manual welding, which significantly improves the efficiency of welding operations.

(3) **Laser processing robot**

Laser processing robots are applied to laser processing. More flexible laser processing operations are achieved through high-precision industrial robots.

(4) **Vacuum robot**

Vacuum robots are the robots that work in a vacuum environment, and are mainly used in the semiconductor industry to realize the transfer of wafers in vacuum chambers.

(5) **Cleaning robot**

Clean robots are the industrial robots used in a clean environment. With the continuous improvement of production technology level, requirements for production environment are becoming increasingly demanding. Many modern industrial products are required to be produced in the clean environment, and clean robots are the key equipment required for production in the clean environment.

High-end equipment manufacturing industry is one of the seven major areas of strategic emerging industries proposed in the national "Twelfth Five-Year Plan". Among them, industrial robots are one of the key directions of high-end equipment manufacturing industry. It vigorously promotes the industrialization of robots and its systems. The development of welding, handling, assembly and other industrial robots and their complete systems are important projects of "Made in China 2025". Intelligent manufacturing is an extension of industrial robot products, is also the integration of various high-tech products in modern production. Industrial robots are the key link to realize the upgrading of intelligent manufacturing equipment, enhance the overall strength of China's manufacturing industry, and truly realize "Intelligent manufacturing". Industrial robots mainly need to solve the following core technologies [8].

(1) **High-precision motion and positioning technology**

Through high-precision sensors and innovative motion mechanism design, it enables the robot to reach the human-level tactile perception array. Through the high-precision hydraulic and the electrical system, it provides industrial robots with the flexible dynamic performance under complex manufacturing environment. In terms of high-precision and high-efficiency of the actuator, the precision, repeatability, resolution and various other properties of industrial robots can be improved by means of improved mechanical devices, selection of advanced materials, and installation of intelligent sensors. Innovative industrial robots such as exoskeletons, intelligent prosthetics and other mechanisms enable industrial robots to achieve higher load ratios, lower emission actuators, and natural interaction mechanisms between humans and machines, etc.

(2) **Autonomous Navigation Technology of Industrial Robots**

Safe autonomous navigation is realized in unstructured environments consisting of static obstacles, vehicles, pedestrians, and animals. For the handling robots loading and unloading raw materials on assembly lines, AGV industrial robots efficiently transporting raw materials to finished products, industrial robots for similar logistics operations for storage and deployment, the robots mining and construction equipment, autonomous navigation is took as the key technology for the further in-depth research and technological development. A typical application is the autonomous navigation of driverless cars, which have been developed to drive in any modern town with clear lighting and road signs, and have been able to demonstrate that it is on a par with manned vehicles in terms of safety.

(3) Environmental perception and sensing technology of industrial robots

In the future, industrial robots will greatly enhance the perception system of factories to detect task progresses of robots and surrounding equipment, detect the production of components and product assemblies in time, and estimate emotional and physical states of production person. It requires high-precision tactile and force sensor and image parsing algorithms. Physical models for assembly tasks and tracking their progress are constructed with high-precision sensors to reduce uncertainty in automated production sessions. Industrial robots for multi-variety, low-volume production will be more flexible and will be able to operate in unstructured environment that include human/producers to participate, which will make it more difficult to perceive and autonomously navigate in the unstructured environment.

(4) Human–computer interaction technology of industrial robots

In the future research and development of industrial robots, the new human–machine cooperation will be emphasized. The fully immersive graphical environment, 3D holographic environment modeling, realistic 3D virtual reality devices and multi-physical effects such as force, temperature and vibration are researched. In order to achieve the goal of harmonious coexistence between robots and human living and behaviors environment, the key issues that need to be solved include: intrinsic safety problems of robots, autonomous adaptation problems; operation problems of diversified operation tools, efficient human–computer collaboration problems. In the production environment, the safety of interaction between human and robots is focused. According to the needs of end users, industrial robot systems, its related products and tasks are designed to ensure natural human–robot interaction.

(5) Open control system of industrial robot based on real-time operating system and high-speed communication bus

Open control system of industrial robots based on real-time operating system and high-speed communication bus adopts the distributed software structure design based on the modular structure to realize the seamless connection between different functions of the robot system. It reduces the difficulty of integrating robot systems by reasonably dividing the robot modules, and improves the real-time performance of software system of robot control system. It overcomes key technologies such as the compatibility of existing robot open source software with robot operating systems, industrial robot modular software and hardware design and interface specifications and software evaluation and testing methods for integrated platforms, openness of software and hardware for industrial robot control systems.

1.2.2 3D Printing

3D printing technology is a type of rapid prototyping manufacturing technology. Its forming principle can be described as Fig. 1.1 [9]:

The process of 3D printing is as follows:

Step 1: 3D modeling

3D modeling can be constructed directly by 3D modeling software, or by product entity scanning, and then reversely obtain the 3D model from data point cloud data.

Step 2: Triangle meshing

According to the quasi-industry standard STL file format in the field of 3D printing, the curved surface of the 3D model needs to be planarized, and the curved surface of the 3D solid is fitted by some smaller triangular patches.

Step 3: Layered processing

In order to obtain the G-code used to drive the 3D printer, it is necessary to be obtain the contour data and filling data of the cross-section, so the STL file produced in the

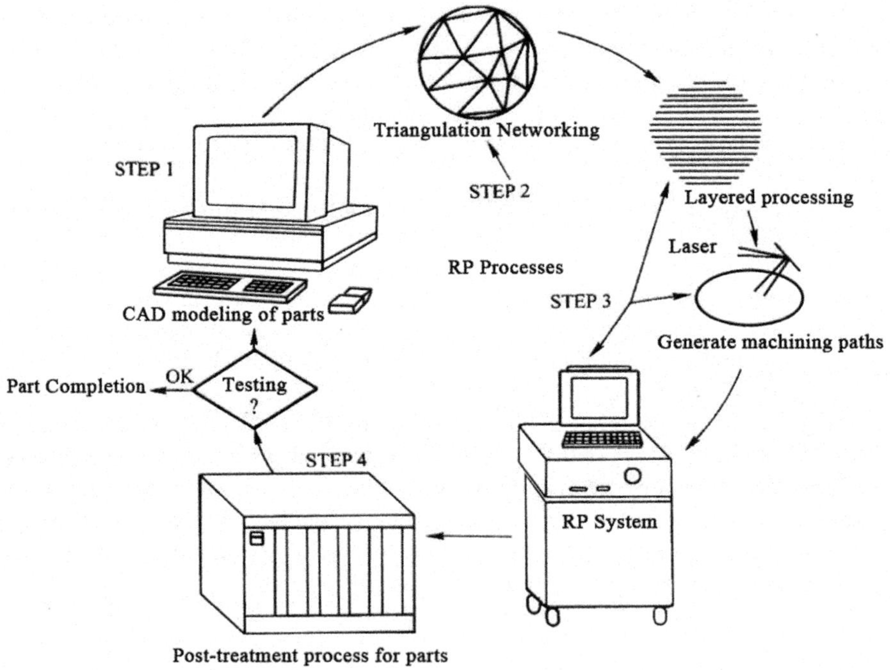

Fig. 1.1 3D printing schematic

second step needs to be layered, and obtain the data after layered processing, such as generate G-code via software.

Step 4: Forming process

The 3D printer prints stacked parts layer by layer according to the G-code in Step 3.

Step 5: Post-processing

Depending on the molding process, the moldings are ground, polished and sintered to obtain the final product.

Due to manufacturing solid parts in a short time, 3D printing technology is called one of the most promising technologies in the twenty-first century. According to the difference between molding materials and molding process principles, 3D printing technologies can be divided into Stereo Lithography Appearance (SLA), Laminated Object Manufacturing (LOM), Fused Deposition Modeling (FDM), Selective Laser Sintering (SLS), Three-Dimensional Printing Method (3DP) [10].

(1) **Stereo Lithography Appearance (SLA)**

The computer-controlled laser beam irradiating the liquid photosensitive resin in the liquid tank. The liquid photosensitive resin rapidly cures to form a layer of contour with the same cross-section as the corresponding 3D model. The worktable descends to the second layer cross section, where the laser beam will cure the photosensitive resin, and the second layer of cured resin will be firmly adhered to the previous layer. The above process is repeated continuously to produce the entire solid part.

(2) **Laminated Object Manufacturing (LOM)**

The thin-shaped material (such as bottom-coated paper rolls or metal foils, etc.) is heated and connected by hot pressing rollers, then is cut into the required layer shape by a CO_2 laser beam. The large-scale models and thick-walled samples that are difficult to manufacture with SLA processes can be formed to facilitate simple design conception and functional analysis. The strength of produced parts is lower.

(3) **Fused Deposition Modeling (FDM)**

The thermoplastics material is extruded by a computer-controlled nozzle according to the level geometric information of the 3D model to form solid samples from the bottom up. The biggest feature of the FDM method is its speed.

(4) **Selective Laser Sintering (SLS)**

The material powder is scanned by the computer-controlled laser according to the cross-sectional geometry information. The scanned powder is first melted and then solidified together. The parts manufactured by the SLS method have high precision and high strength, so samples can be used for functional tests or assembly simulations.

(5) **Three-dimensional printing method (3DP)**

The powder layer is selectively sprayed with the binder according to the cross-sectional shape of the part. Then, another layer of powder material is sprayed. The above steps are repeated. Finally, the finished part is put into the furnace for sintering to strengthen. The parts manufactured by the 3DP method have high strength. This method usually can be used for mold manufacturing.

1.2.3 RFID

Radio Frequency Identification (RFID) is a non-contact automatic identification technology. Its basic principle is to use the transmission characteristics of radio frequency signals or spatial coupling to realize automatic identification of objects or commodities. In 1948, Harry Stockman published the article "Communication Using Reflected Power", which laid the theoretical foundation of RFID. RFID system generally consists of three parts: tag, reader and central processing unit. The tag consists of coupled antennas and chips. Each tag has a unique Electronic Product Code (EPC), and are attached to the marked object. The reader is used to read and write tag information, and its external antenna can be used to send and receive radio frequency signals. The central processing unit includes middleware and database, which are used to process the read and write tag information, its system composition framework is shown in Fig. 1.2.

Compared with other automatic identification technologies, this technology has the advantages of strong anti-interference ability, large amount of information, non-visual reading and writing and long life. It is widely used in logistics, supply chain, animal and vehicle identification, access control system, library management, automatic charging and manufacturing and other fields, and is known as one of the most promising information technologies in the twenty-first century [11]. The combination of RFID technology and manufacturing technology can realize the automation and real-time collection of various production data, and grasp timely the production

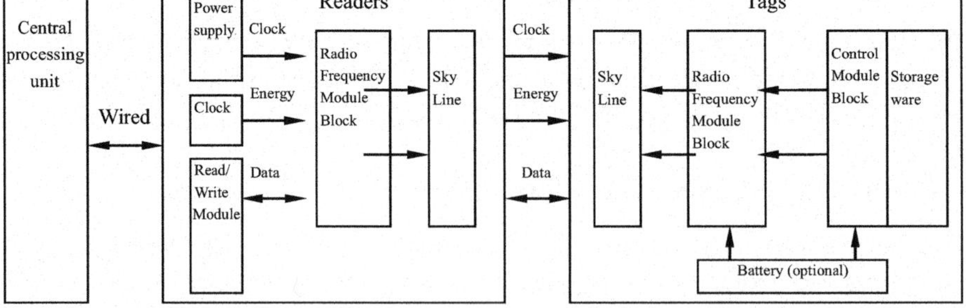

Fig. 1.2 Technical principle

plan and the running status of MES. It can effectively track, manage and control the resources and work-in-process required for production, and realize the visual management of the production process. It can enhance the accuracy and timeliness of material scheduling at production sites, strengthen process monitoring, and provide the overall operating efficiency of MES.

In recent years, the development of RFID technology and its application in manufacturing have received double attention from academia and industry. Dalian Machine Tool (CNC) Co., Ltd. manages tool using the RFID system in the workshop to efficiently improve the efficiency of tool use. Qingdao Haier begun to use industrial-grade readers for product collection's data on refrigerator production line. Wuhan Huaweike Intelligent Science and Technology Limited Company develops special RFID tags such as high temperature and high pressure resistance, long life, anti-metal, laser holographic RFID tags to meet the application requirement in manufacturing, anti-counterfeiting, oil exploration. Meanwhile it has been successfully applied in the manufacturing process management of domestic leading enterprises such as SANY Heavy Industry and Midea.

1.2.4 Wireless Sensor Network

WSN is a wireless network composed of a large number of stationary or mobile sensors in a self-organizing and multi-hop manner. It cooperatively sense, collect, process and transmit information in the geographical area covered by the network, and ultimately send this information to the owner of the network. Wireless sensor networks have many types of sensors that can detect a variety of phenomena in the surrounding environment, including earthquake, electromagnetism, temperature, humidity, noise, light intensity, pressure, soil composition, size, speed, and direction of moving objects. Wireless sensor networks have the following characteristics:

(1) **Large-scale**

The large-scale nature of sensor networks includes two meanings: on the one hand, sensor nodes are distributed in a large geographical area. For example, the use of sensor networks for forest fire prevention and environmental monitoring in large pristine forests, where requires the deployment of a large number of sensors. On the other hand, sensor nodes are deployed very densely and a large number of sensor nodes are densely deployed in a relatively small space.

(2) **Self-organization**

Sensor nodes in the sensor network have the ability of self-organization, can automatically configure and manage, and automatically form a multi-hop wireless network system that forwards monitoring data through topology control mechanisms and network protocols.

(3) **Dynamicity**

The topology of sensor networks may change due to the following factors: (1) Sensor node malfunction or failure caused by environmental factors or power depletion; (2) Changes in environmental conditions may cause changes in the bandwidth of wireless communication links, even intermittently on and off; (3) The three elements of sensor network, sensor, sensing object and observer may have mobility; (4) The addition of new nodes. Sensor network systems are required to be able to adapt to such changes.

(4) **Integration**

Sensor nodes have low power consumption, small size and low price.

1.2.5 IoT and Cyber-Physical Fusion System

In 1999, MIT Auto ID Center gave an earlier definition of "Internet of Things" as a network covering everything in the world (Internet of Things) on the basis of the computer Internet, using RFID, wireless data communication and other technologies to realize the automatic identification of items and the interconnection of information sharing [12]. In 2005, the "ITU Internet Report 2005: Internet of Things" [13] issued by the International Telecommunication Union (ITU) formally defined the Internet of Things as: though radio frequency identification (RFID), infrared sensors, global positioning systems, laser scanners and other information sensing equipment, connect any item with the Internet according to the agreed protocol to exchange and communicate information, in order to realize Intelligent identification, positioning, tracking, monitoring and management of network. At the same time, it is pointed out that the Internet of Things is an extension of Internet applications, and "RFID, sensor technology, nanotechnology, and intelligent embedded technology" will be the four core technologies for realizing the Internet of Things. Although the concept of the Internet of Things has been developed for more than ten years, there is still no clear and unified definition. In a CERP-IoT SRA [14] published by the European Union in September 2009, the "Internet of Things" is defined as: The Internet of Things will be an integral part of the future internet, and which is a dynamic global network architecture with self-organization capabilities based on certain standards and interoperable communication protocols. Both physical and virtual "things" have identity, physical attributes and virtual characteristics, which can be seamlessly linked to the information network by intelligent interfaces. The definition in China [15, 16] is: "Internet of things, refers to the huge network formed using the various information sensing devices, such as radio frequency identification devices, infrared sensors, global positioning systems, laser scanning and other devices combined with the Internet. The purpose of network is to connect all items to facilitate identification and management. It is believed that the Internet of Things should have three characteristics: comprehensive perception, reliable transmission and intelligent processing.

The object dynamics is perceived using various available sensing methods in real-time at any time. Through the integration of various information networks and the Internet, the perceived information is transmitted accurately and reliably in real time. Finally, the massive amounts of data and information are analyzed and processed to intelligently control objects using cloud computing and other intelligent computing technologies.

Similar to the Internet of Things, the Cyber-Physical System (CPS), also known as a "virtual network, physical entity" production system, aims to make the physical system capable of computing, communication, precise control, remote cooperation and autonomy. Through the Internet, various corresponding autonomous control systems and information service systems are formed to complete the organic coordination between the real society and the virtual space. Compared with the Internet of Things, CPS emphasizes on loop feedback, requiring systems to be able to sense the physical world and provide feedback to the physical world through communication and computation. In such a system, a part can figure out which services it needs. The gradual upgrade of existing production facilities through digitalization leads to a completely new architecture. CPS is a multi-dimensional complex system that integrates computing, network and physical environment. Through the organic integration and deep collaboration of 3C (Computation, Communication, Control) technologies, it realizes real-time perception, dynamic control and information services for manufacturing. CPS realizes the integrated design of computing, communication and physical systems, which can make the system more reliable, efficiency, and real-time collaboration, and has important and broad application prospects. The CPS system deeply embeds computing and communication into the physical process, and makes it closely interact with the physical process.

1.2.6 Industrial Big Data

Big data generally refers to data sets that are particularly large in volume and data category, which cannot be captured, managed and processed by conventional software tools within a certain time range. It requires a new processing mode to have stronger decision-making capacity, insight and process optimization capabilities of the massive, high-growth and diverse information assets. As early as 1980, the futurist Toffler, in his book "The Third Wave", enthusiastically praised "big data" as "the cadenza of the third wave" [17]. After the top journal "Nature" launched the cover column of "big data" in 2008, the term "big data" gradually became a buzzword [18]. The concept of big data contains at least three meanings, namely "large volume of data", "large data type", and "large data storage range". Different research institutions have provided different definitions for the meaning of big data. McKinsey Global Institute (MGI) described Big Data technology as the collection of data that cannot be captured, managed and processed by traditional database software tools within a certain period of time [19]. The National Institute of Standards and Technology (NIST) defines Big Data as large data, rapidly accessible, or various forms, which are difficult to be effectively analyzed by traditional relational data analysis methods, or massive horizontal scaling to process efficiently [20].

Big data has five main technical characteristics, namely the 5 V characteristics of Big Data (IBM).

(1) Volume (high volume): Large data volumes, with units of measurement rising from the TB level to the PB, YB and above levels
(2) Velocity (high speed): in the case of very large data volumes, it also maintains high-speed, real-time data processing
(3) Variety: rich and diverse data types, including production logs, pictures, sounds, videos, location information and other diverse and multi-dimensional information
(4) Value (low value density): there is massive data information, but there is a lot of irrelevant information. It is necessary to use artificial intelligence technology for data analysis and mining
(5) Veracity: When converting from the real world, the logical world, and the data world, the original information of the real world is basically maintained.

The generation, analysis and use of big data can be illustrated by Fig. 1.3.

As shown in Fig. 1.3, big data really serves customers and needs a series of pre-processing work. This pre-processing process is the process of data analysis and mining through artificial intelligence technology. Due to the broad application prospects of big data technology, many more mature big data cloud platforms have emerged so far. Three stages of industrial big data development are shown in Table 1.1.

For the manufacturing industry, big data comes from all stages of the manufacturing lifecycle, and includes three main parts: (1) Raw data from supply chain

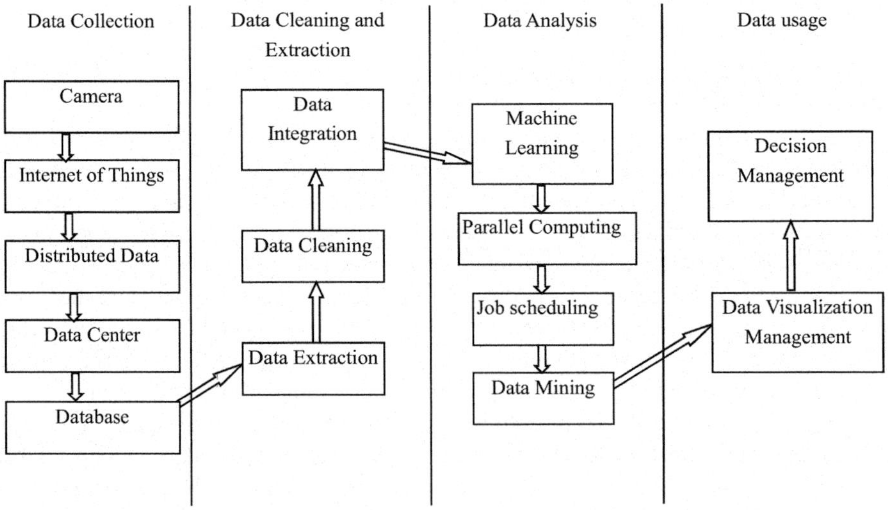

Fig. 1.3 4 stages of big data technology

Table 1.1 Three stages of industrial big data development

	Phase I	Phase II	Phase III
Time	1990–2000	2000–2010	2010-present
Core Technologies	Remote data monitoring, data acquisition and management	Big data center and data analysis software	Data analysis platform and advanced data analysis tools
Problem Object/Value	Product-centric status monitoring	Use-focused information services	User-centric platform-based services
Business Model	Product-based additional services	Product leasing system and long-term service contracts	On-demand personalized self-service model, sharing economy
Representative companies and technology products	GM Onstar™, OTIS REMTM, GE Medical InSite	GE Aviation On-wing Support TM, Komatsu Komtrax™, Alstom Track Tracer™	IMS NI LabVIEW based Watchdog Agent, GE Predix Platform

management system (SCM), enterprise resource planning system (ERP), and manufacturing execution system (MES). (2) Dynamic manufacturing data from the MES process collected by sensors, RFID, wireless Wi-Fi. (3) From external manufacturing data, such as feedback data from other regions or national manufacturing systems, customer demand data collected from the Internet. Using the data from the above three aspects, a manufacturing data analysis model based on big data can be established, which can configure and optimize manufacturing resources in an active, all-round and multi-dimensional manner. The data in the manufacturing process comes from different spatial geographical locations. The production process data and operational monitoring data can be collected through the Internet, Wi-Fi, sensors and other means. After data cleaning and data aggregation, the data is transformed into useful data by data analysis and mining technology that drives the operation of manufacturing systems. Each enterprise system will also form its own manufacturing data to realize the application of big data through the manufacturing cloud platform.

1.2.7 Cloud Computing Technology

The idea of cloud computing can be traced back to the 1960s, John McCarthy mentioned that "sooner or later, computing will become a kind of public infrastructure" [21]. After IBM and Google announced their cooperation in the field of cloud computing in October 2007 [22], cloud computing attracted the attention of many people and quickly became a research hotspot in both industry and academia. The description of cloud computing in IBM's technical white paper "Cloud Computing" is as follows: Cloud computing refers to a system platform or a type of application. Cloud computing platform can dynamically deploy, configure, reconfigure and cancel services on demand. It can use either physical servers or virtual servers. It also includes storage area networks, network devices, firewalls and other security devices. Cloud computing is also an Internet-accessible application, a web application or web service that runs in large-scale data centers and powerful servers. It allows any user to access the cloud computing platform through appropriate Internet equipment and security rules.

The definition of cloud computing from Baidu Encyclopedia: cloud computing is an Internet-based model for the addition, use and delivery of related services. It usually involves the provision of dynamic, easily scalable and often virtualized resources through the Internet. Cloud computing allows you to experience the power of 10 trillion operations per second. Users access data centers and perform computing on their own demand through computers, laptops, mobile phones and other means [23].

China Cloud Computing Network defines cloud computing as: cloud computing is the development of Distributed Computing, Parallel Computing and Grid Computing, or the commercial realization of these scientific concepts [24].

Professor Luo Junzhou of Southeast University summarizes the characteristics of cloud computing as follows [25]:

(1) **Flexible service**

Services can be quickly scaled to automatically adapt to dynamic changes in business load. The resources used by users are consistent with business requirements. The flexible service avoids the degradation of service quality caused by overloaded or redundant server performance.

(2) **Resource pooling**

Resources are managed uniformly in a shared resource pool. Using virtualization technology, resources are shared among different users. The resource placement, management and allocation strategies are transparent to users.

(3) **On-demand service**

Resources such as applications, data storage and infrastructure are provided to users in the form of services. The resources can be allocated automatically based on user demand without the intervention of system administrators.

(4) **Services can be charged**

Monitor users' resource usage and charge for services based on resource usage.

(5) **Ubiquitous access**

Via the Internet users can access cloud computing services using various terminal unit anytime, anywhere.

Based on the above characteristics, the architecture of cloud computing is shown in the following Fig. 1.4.

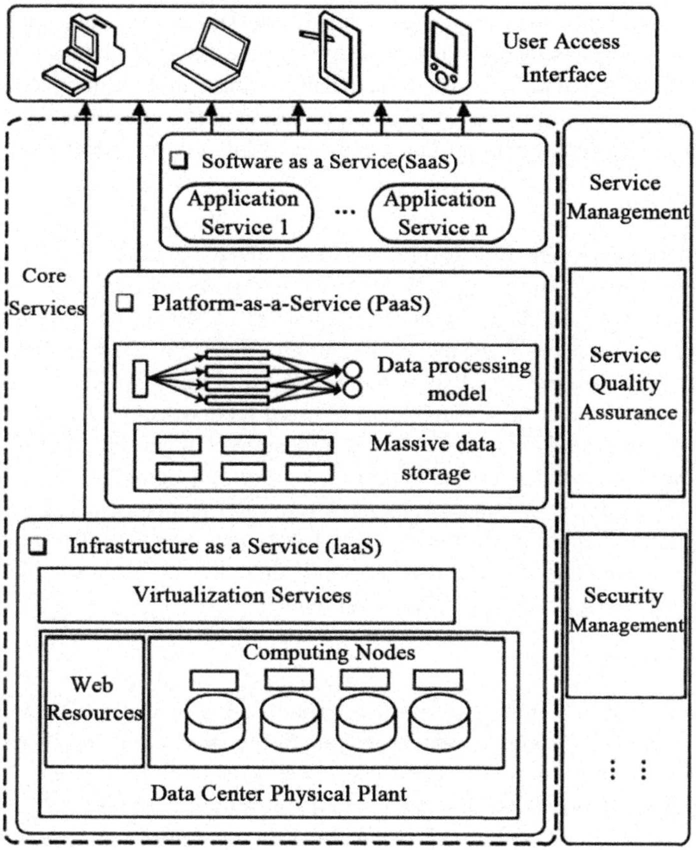

Fig. 1.4 Cloud computing architecture

1.2.8 Virtual Reality

VR adopts computer technology as the core of advanced technology to generate a realistic visual, auditory and tactile virtual environment. Users can interact with objects in the virtual environment through necessary input and output devices, and affect each other, and then get the immersive feeling and experience. This computer-generated virtual environment can be a representation of a particular objective world or a purely fictional world [26]. Virtual reality technology integrates various advanced technologies such as computer simulation technology, computer-aided design and graphics, multimedia technology, artificial intelligence, network technology, sensing technology, real-time computing technology, and psycho-behavioral research. It has greatly facilitated people's exploration of the macroscopic world. In the virtual reality environment, participants interact with objects in the virtual environment with the help of virtual reality interaction devices such as data gloves, 3D mouse orientation trackers, joysticks, helmet-mounted displays, headsets, and data suits. Objects in virtual reality provide real-time feedback, resulting in immersive, interactive visual simulation and information exchange. Immersion, interactivity and real-time performances are the most important features of virtual reality technology [27]:

(1) **Immersion**

In the virtual environment, the designer observes and experiences the design process and design results through a variety of perceptual pathways, including the stereoscopic display with depth perception, fine-grained 3D sound and tactile feedback. The ability of visualization in the virtual environment has been further enhanced. With the help of new graphic display technology, designers can get real-time, high-quality, and depth-perceptive stereoscopic visual feedback. On the other hand, the 3D sound in the virtual environment enables designers to get the more accurate feel the position of the object. Haptic feedback supports designers to directly feel the reaction force of objects when grasping and moving them in the virtual environment. Under the combined effect of multi-sensing forms, users can be fully "immersed" in the virtual environment. They experience and perceive the virtual world in multiple ways, multiple perspectives, and authentic manner.

(2) **Interactivity**

The human–computer interaction in the virtual reality system is a kind of almost natural interaction, in which users can operate the objects in the virtual environment through their own natural skills such as language, body movement or action. The computer adjusts the images and sounds presented by the system in real time according to the user's body movements and language information. Users can use different interactive means to complete the same interactive task. For example, when performing part positioning operations, designers can express the positioning coordinates of the part by voice commands, or by dragging the part to the positioning point through gestures. Various interactive methods have their own advantages in terms of information input. The advantage of voice is that it is not limited by space, and the designer can manipulate the design object without "touching" it. The advantage of direct 3D operations such as gestures is the directness of the motion control. The combination of multiple means of interaction increases the bandwidth of information input, and is helpful for the effective communication of interactive intentions.

(3) **Real-time performance**

Two important indicators of dynamic characteristics and interaction delay characteristic are used to measure the real-time performance of the virtual reality system. In visual terms, it is required to generate and display 30 frames graphics per second, otherwise it will produce a sense of discontinuity and beating. In tactile terms, to achieve the force sensation of virtual reality, the contact force must be calculated and updated at a speed of 1000 frames per second. For the interaction action generated by the human, the system should immediately react and generate the corresponding environment and scenarios, the time delay between them should not be greater than 0.1 s.

1.2.9 Artificial Intelligence

Artificial intelligence (AI) mainly studies the use of artificial methods and technologies to imitate, extend and expand human intelligence and realize machine intelligence. In 1956, four young scholars including McCarthy John, organized a summer symposium on the machine simulation of human intelligence. At this conference, McCarthy proposed the name of "artificial intelligence" for the first time, defined as the science and engineering for manufacturing intelligent machines. Artificial intelligence technology is an emerging marginal discipline, and an interdisciplinary subject in natural and social science. It absorbs the latest achievements of natural and social science, takes intelligence as the core, and forms a new system with its own research characteristics. The field of artificial intelligence research includes knowledge representation, search techniques, machine learning, various methods for solving data and knowledge uncertainty problems. Application areas of AI include expert systems, games, theorem proof, natural language understanding, image understanding and robotics. Artificial intelligence is also a comprehensive discipline, born on the basis of cybernetics, information theory and systems theory, involving philosophy, psychology, cognitive science, computer science, teaching, and various engineering methods.

In recent years, with the advancement of deep learning algorithms and brain-computer interface technology, the research on the basic theories and methods of artificial intelligence has undergone new changes. In 2016, Google artificial intelligence AlphaGo was marked by a 4:1 victory over South Korean chess player Lee Sedol, and artificial intelligence was once again become the focus of public attention. The essence of the technology is big data + deep learning. Through a large amount of training data, a neural network is trained to evaluate a large number of selected points in the situation, and a strategy neural network is trained to be responsible for the moves, using these two networks are used simultaneously in the Monte Carlo search.

The driver assistance system is currently the hottest direction in the field of automotive artificial intelligence. At the perception level, it uses machine vision and speech recognition technology to perceive the driving environment, identify the people in the car, and understand the needs of passengers; at the decision-making level, it uses machine learning models and deep learning models to establish a driving decision-making system that can automatically make judgments. According to the degree of machine intervention, the driverless system can be divided into five stages: no automatic driving (L0), driving assistance (L1), partial automatic driving (L2), conditional automatic (L3) and fully automatic (L4). At present, the technology as a whole is in the transition stage from the fusion control of multiple driver assistance systems, monitoring road conditions and intervening in emergency situations (L2) to the basic realization of automatic driving functions (L3).

Tan Tieniu's team of the Institute of Automation of the Chinese Academy of Sciences insists on systematic innovation from the source of iris image information acquisition in the field of iris recognition, and comprehensively breaks through a series of key technologies such as imaging devices, image processing, feature extraction, recognition retrieval, security and anti-counterfeiting in the field of iris recognition. The computational theory and method system of the iris recognition comparison system has been established, and the largest shared iris image database in the world has been built, which has been used for coal mine personal identification and Beijing city railway monitoring on a large scale, and it has been promoted and used by more than 3000 scientific research teams in 70 countries and regions, which has effectively promoted the development of iris recognition disciplines.

1.3 Architecture of Intelligent Manufacturing System

The intelligent manufacturing system is an intelligent manufacturing network of virtual reality. Its system architecture is the basis for the research, development and application of intelligent manufacturing systems, and is the skeleton and soul of intelligent manufacturing. When building the reference architecture of the intelligent manufacturing system, it is necessary to fully reflect the hierarchical functions of the manufacturing enterprise. The IEC/ISO62264 "Enterprise Control System Integration" standard jointly formulated by the International Organization for Standardization (ISO) and the International Electrotechnical Association (IEC) proposes a functional hierarchy model of manufacturing enterprises, which divides the functions of manufacturing enterprises into 5 levels: Level 0 is the physical processing layer; Layer 1 is the production process perception and control layer; Layer 2 is the monitoring and control layer of the production process; Layer 3 is the manufacturing execution control layer; and layer 4 is the business planning and logistics management [28]. On December 29, 2015, in China, the Ministry of Industry and Information Technology and the National Standard Commission jointly issued the "Guidelines for the Construction of the National Intelligent Manufacturing Standard System" (2015 edition), proposing that there are five layers of Intelligent manufacturing system from bottom to top, namely the equipment layer, the control layer, and the workshop layer, the enterprise layer and the collaboration layer.

The intelligent manufacturing system is based on Product Lifecycle Management (PLM) as the core to form a value-added chain to realize the intelligence of product research and development, production and service. The interconnection and integration between equipment and equipment, between equipment and control systems, and between enterprises and enterprises are realized to establish an intelligent manufacturing enterprise value-added network, which have the characteristics of high flexibility and sustainable optimization. Based on the management model of the whole product life cycle, a cloud-based security network (Internet, mobile Internet, Internet of Things and wireless network) should be built. Based on the analysis, mining, evaluation, prediction and optimization of massive manufacturing data information using big data and cloud computing in the product production and manufacturing process, the horizontal integration, vertical integration and end-to-end integration of system intelligent manufacturing should be realized to build the intelligent function system of manufacturing enterprises with information fusion [29]. Combined with the IEC/ISO 62264 "Enterprise Control System Integration" standard, the research results of experts in the field of digitization and intelligent manufacturing, and the "Guidelines for the Construction of National Intelligent Manufacturing Standard System Construction" (2015 edition), the overall model for an intelligent manufacturing system involves three aspects: system level, full life cycle management and smart features, which is shown in Fig. 1.5.

The overall model of the intelligent manufacturing system is a mapping of the internal elements of intelligent manufacturing and the relationship between them. According to the overall model of the intelligent manufacturing system in Fig. 1.5 the architecture of the intelligent manufacturing system and the components of each level are shown in Fig. 1.6 [29].

The equipment layer is mainly composed of various sensors, intelligent meters, barcodes and other elements such as intelligent equipment, industrial robots, intelligent logistics equipment, intelligent testing equipment, automated guided vehicle(AGV), automatic storage & retrieval system (AS/RS), radio frequency identification (RFID), micro electro mechanical systems (MEMS), and through wireless networks, Internet of Things and other information and communication technologies to solve machine-to-machine communication (M2M), to meet the equipment production process monitoring, command and dispatch, remote data acquisition

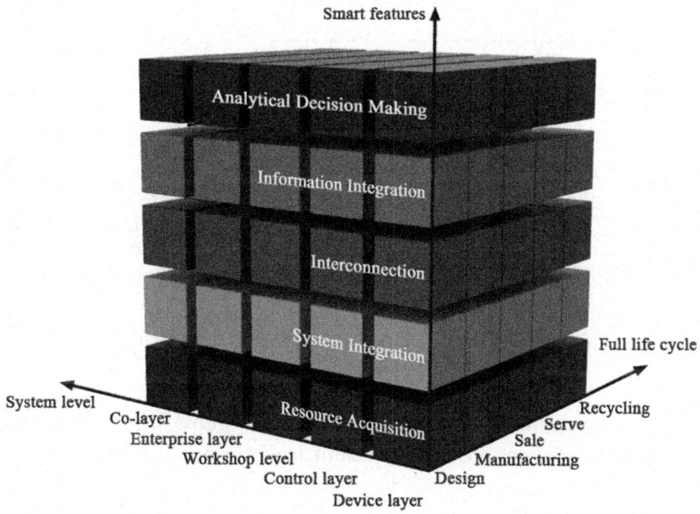

Fig. 1.5 Overall model of intelligent manufacturing system

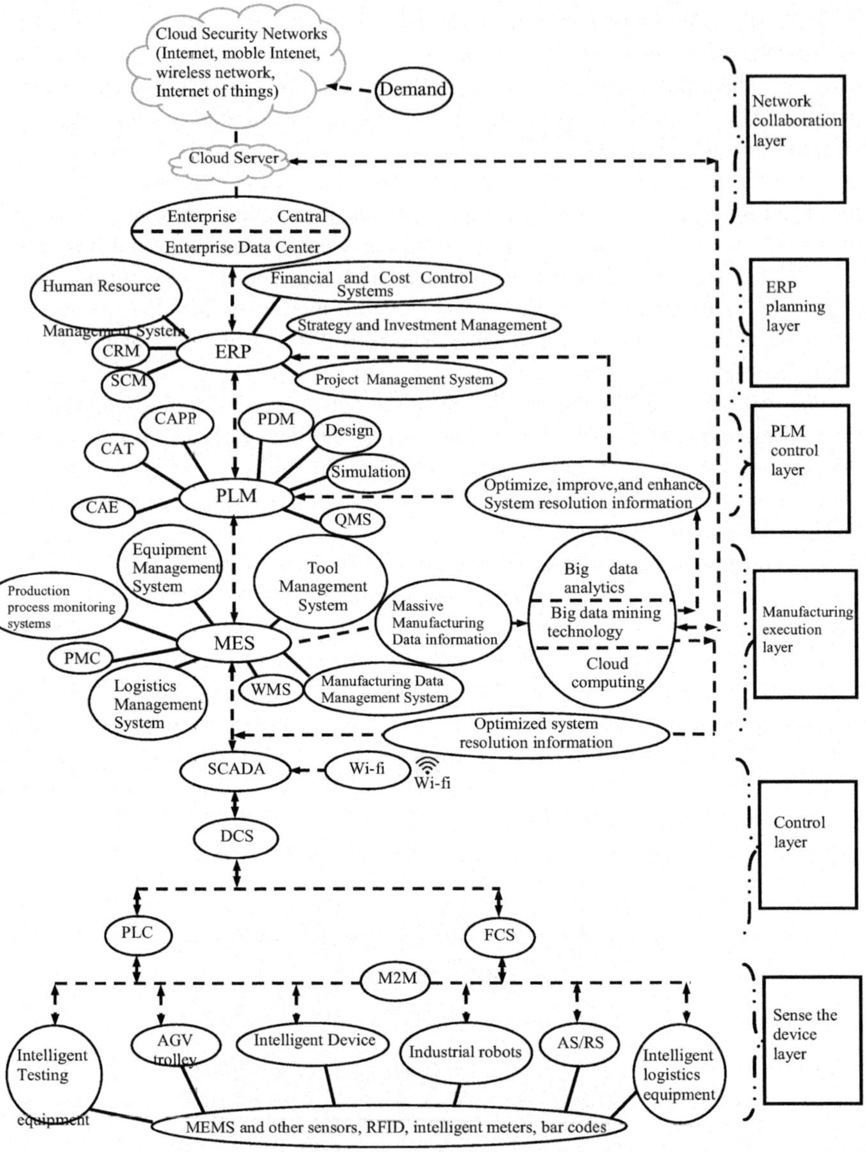

Fig. 1.6 Elements at all levels of intelligent manufacturing system architecture

and measurement, remote diagnosis and other intelligent, digitization and information requirements, and to achieve equipment interconnection intelligence and state perception.

The control layer is mainly composed of Supervisory Control And Data Acquisition (SCADA), Distributed Control System (DCS), Programmable Logic Controller

(PLC) and Fieldbus Control System (FCS), and other elements. Combined with the wireless network control command information and the command information analyzed, mined, evaluated, predicted and optimized by big data and cloud computing, the monitoring of the sensing device layer equipment and the collection of manufacturing data information is realized.

The workshop layer mainly refers to the Manufacturing Execution System (MES), which is mainly composed of equipment management system, tool and tooling management system, logistics management system, manufacturing data management system, production process monitoring system, Warehouse Management System (WMS) and Product material control (PMC) and other elements. According to the collaborative production of different enterprises in the network collaboration layer industry chain and the product manufacturing process optimized by the PLM management and control layer, we can control the status of the product production process in real time, and quickly deal with material shortages, equipment failures, and personnel absenteeism and various abnormal situations in the manufacturing process, and analyze, mine, evaluate, predict and optimize the massive data, information, and charts formed in the product manufacturing process by big data and cloud computing in real time, so as to be presented on the visual interface of PLM management and control layer, the ERP planning layer and collaborative enterprises, and independently optimize the execution of the product manufacturing process, so as to achieve plans of the scheduling and producing intelligence, production processes of the coordination intelligence, decisions of the supporting intelligence, production resources of the management and control intelligence, and quality processes of the control intelligence, and real-time analysis, independent decision-making, and accurate execution.

The enterprise layer includes the Enterprise Resource Planning (ERP) system and the Product Lifecycle Management (PLM) system. The ERP system consists of the financial and cost control system, the strategy and investment management, the human resource management system, the project management system, Supply Chain Management (SCM) and Customer Relationship Management (CRM) and other elements. According to the collaborative research and development, production, and service division of different enterprises in the industrial chain of the network collaboration layer, combined with product databases, big data, and cloud computing to analyze, mine, evaluate, predict, and optimize the massive manufacturing data information in the Intelligent manufacturing process, rationally allocate and optimize the internal resources (people, financial, material, information) of enterprises, reduce the cost of resource occupation, and realize planning and decision support intelligence.

The PLM system is mainly composed of design, simulation, Computer Aided Process Planning (CAPP), Product Data Management (PDM), Quality Management System (QMS), Computer Aided Testing (CAT) and Computer Aided Engineering (CAE) and other elements. According to the collaborative research and development, the production division and the resource allocation of the ERP planning layer of different enterprises in the industrial chain of the network collaboration layer, combined with product database, big data, cloud computing to analyze, mine, evaluate, predict and optimize the massive manufacturing data information in the Intelligent manufacturing process, and collaborate with enterprises to optimize design, virtual testing, production and manufacturing process simulation, evaluation, control, and production and manufacturing data sharing to realize Intelligent decision-making in the product development process.

The network collaboration layer is mainly composed of cloud security network (Internet, mobile Internet, Internet of Things and wireless network), cloud server, enterprise central server and enterprise data center and other elements, realizing collaborative research and development, Intelligent production, accurate logistics, real-time data analysis, calculation and services of different enterprises in the industry chain through cloud security network to share information.

References

1. Gonzalez Aguirre JA, Osorio Oliveros R, Rodríguez Hernández KL, Lizárraga Iturralde J, Morales Menendez R, Ramírez Mendoza RA, Ramírez Moreno MA, Lozoya Santos JDJ (2021) Service robots: trends and technology. Appl Sci 11(22)
2. Kim SH, Nam E, Ha TI, Hwang SH, Lee JH, Park S-H, Mi B-W (2019) Robotic machining: a review of recent progress. Int J Precis Eng Manufact 20(9)
3. Painter CW (1987) Defining the industrial robot. Prod Eng 66(9)
4. Wang F (2022) Review of the application status and development trend of industrial robots. In: SHS web of conferences, p 144
5. Zhu Y, Zhang K (2021) The application status and development prospects of artificial intelligence and industrial robots in the manufacturing industry. J Phys Conf Ser 2066(1)
6. Li Y (2022) Analysis of the application of industrial robots in intelligent manufacturing. Int J Front Eng Technol 4(7)
7. Zhou Lianyang, Wang Fei, Wang Ningwei, Yuan Tongwen (2021) Application of industrial robots in automated production lines under the background of intelligent manufacturing. J Phys Conf Ser 1992(4)
8. Reshetnikova MS, Pugacheva IA, Lukina YD (2021) The evolution of robotics in China: accumulated experience and prospects. In: SHS web of conferences, p 114
9. Huzaim NHM, Rahim SZA, Musa L, Abdellah AE, Abdullah MM, Bakri A, Rennie A, Rahman R, Garus S, Błoch K, Sandu AV, Vizureanu P, Nabiałek M (2022) Potential of rapid tooling in rapid heat cycle molding: a review. Materials 15(10)
10. Li JZ, Alkahari MR, Rosli NAB, Hasan R, Sudin MN, Ramli FR (2019) Review of wire Arc additive manufacturing for 3D metal printing. Int J Autom Technol 13(3)
11. Zhang M, Liu Z, Shen C, Wu J, Zhao A (2022) A review of radio frequency identification sensing systems for structural health monitoring. Materials 15(21)
12. Sanjay S, David LB, Kevin A (2000) MIT Auto ID WH-001: the networked physical world. MIT Press, Massachusetts
13. International Telecommunication Union (ITU) (2005) ITU Internet reports 2005: the internet of things. World Summit on the Information Society (WSIS), Tunis
14. Prakash V, Savaglio C, Garg L, Bawa S, Spezzano G (2022) Cloud- and Edge-based ERP systems for industrial Internet of Things and intelligent factory. Procedia Comput Sci 200
15. Mirani AA, Velasco Hernandez G, Awasthi A, Walsh J (2022) key challenges and emerging technologies in industrial IoT architectures: a review. Sensors 22(15)
16. Zarei M, Mohammadian A, Ghasemi R (2016) Internet of things in industries: a survey for sustainable development. Int J Innov Sustain Dev 10(4)

17. Xie S, Zhang W, Xue F, Li D, Liu Y, Fleischer J, Ehrmann C (2022) Industry 4.0-oriented turnkey project: rapid configuration and intelligent operation of manufacturing systems. Machines 10(11)
18. Abdalla HB (2022) A brief survey on big data: technologies, terminologies and data-intensive applications. J Big Data 9(1)
19. Chu X, Nazir S, Wang K, Leng Z, Khalil W (2021) Big data and its V's with IoT to develop sustainability. Sci Prog 2021
20. Marwane S, Mostafa H (2022) Typical application of big data technology in different fields. J Res Sci Eng 4(9)
21. Lenhard J (2014) Disciplines, models, and computers: the path to computational quantum chemistry. Stud Hist Philos Sci 48
22. Cloud Computing Stack Layers Market (2017) Global key players—salesforce, IBM, Google, SAP, H&P Helion analysis and forecast to 2022. M2 Presswire
23. Rahman MA, Shakur MS, Ahamed MS, Hasan S, Rashid AA, Islam MA, Haque MSS, Ahmed A (2022) A cloud-based cyber-physical system with industry 4.0: remote and digitized additive manufacturing. Automation 3(3)
24. Agbaegbu JB, Arogundade OTe, Misra S, Damaševičius R (2021) Ontologies in cloud computing—review and future directions. Future Internet 13(12)
25. Jawed SM, Sajid M (2022) A comprehensive survey on cloud computing: architecture, tools, technologies, and open issues. Int J Cloud Appl Comput (IJCAC) 12(1)
26. de Sá AG, Zachmann G (1999) Virtual reality as a tool for verification of assembly and maintenance processes. Comput Graph 23(3)
27. Zhu W, Liu Y, Wang Y (2022) A complex machinery design-oriented digital prototype model driven method. J Northwest Polytechn Univ 40(4):901–908 (in Chinese)
28. Ministry of Industry and Information Technology (2018) Standardization administration of people's Republic of China. In: Guide for the construction of national intelligent manufacturing standard system (2018 Edition). China Standards Press, Beijing
29. Suvarna M, Yap KS, Yang W et al (2021) Cyber-physical production systems for data-driven, decentralized, and secure manufacturing—a perspective. Engineering 10

Chapter 2
Intelligent Optimization of Cutting Parameters

2.1 Introduction

SiCp/Al composites are widely applied in the aerospace, high-speed rail, and deep-ocean engineering fields [1, 2]. However, the high hardness of SiC particles makes it difficult to obtain good surface finish and high machining efficiency when cutting SiCp/Al, which also tends to result in severe tool wear. Research on cutting SiCp/Al composites started in 1985 [3]. To date, many research articles have been published that cover the deformation mechanism and surface finish of SiCp/Al composites, tool wear, cutting force, etc. [4–6]. In the past thirty years many works have been done to improve the machinability of SiCp/Al composites, covering the deformation of SiCp/Al composites, surface finish, tool wear, cutting force [6–9]. But most of these researches focused on the low-volume fraction SiCp/Al composites (< 40 vol.%). It has been proved that the volume fraction has a crucial impact on the machinability, especially, on surface roughness, cutting force and tool wear [10–12]. The difficulties and cost of manufacturing SiCp/Al composites increase dramatically with the increased volume fraction of SiC particles while the percentage of SiC particle volume exceeds 40% [13, 14]. Limited by the manufacturing technology of mixing a large amount of SiC particles into aluminium matrices, the investigations of machining high-volume fractions SiCp/Al composites were not reported until 2012 [15].

2.1.1 The Study of Surface Roughness

Surface roughness plays an important role in the evaluation of machining performance. The production of good surface roughness is essential for many components [16, 17]. However, the presence of hard SiC particles makes it extremely difficult for SiCp/Al composites to obtain good surface roughness. There are a vast number

J. Wang, *Intelligent Manufacturing System and Intelligent Workshop*,
Advanced and Intelligent Manufacturing in China,
https://doi.org/10.1007/978-981-99-2011-2_2

of defects on the machined surface of an SiCp/Al composite, such as voids around the SiC particles and SiC fractures. In particular, when the volume fraction of SiC is relatively large, the machined accuracy is supposed to deteriorate greatly [18].

Published literature indicates that experimental investigations are the main research approach for surface roughness. Surface roughness values can be measured when cutting experiments are carried out. Based on experimental roughness data, the significance of various parameters has been analysed, and the optimal roughness results have been predicted using scientific methodology. Seeman et al. [19] applied a machinability evaluation model to 20% SiCp/Al composites and investigated the influences of the cutting speed (v_c), feed rate (f), cutting depth (a_p) and machining time on surface roughness. The cutting parameters were optimised for the optimum roughness value using response surface methodology. Kumar et al. [20] investigated the effects of v_c, a_p, and f in the machining of 10% SiCp composites. The surface roughness obtained using a tungsten carbide tool was compared with that obtained using polycrystalline diamond (PCD) inserts. Ekici et al. [21] investigated the influence of v_c,f and the coating type on the surface roughness in machined 10% SiCp/Al composites. Response surface methodology (RSM) was used to optimise the cutting parameters. Tamang et al. [22] predicted the surface roughness in the machining of SiCp/Al composites using RSM and an artificial neural network (ANN).

However, these studies investigated the surface roughness obtained from machining low-volume-fraction SiCp/Al composites. Moreover, in most research, the tool nose radius (r_ε) is not considered. At the finishing stage of machining, a small a_p, a small f, and a high v_c tend to be selected to achieve small surface roughness values. The influence of the tool nose radius must be considered due to te small depth of cut, especially when this value is close to the tool nose radius [23–25]. Baburaj et al. [26] modelled surface roughness by considering the effect of the nose radius. However, this investigation was developed for low-volume-fraction SiCp/Al composites (15%).

2.1.2 *The Study of the Modeling of Cutting Forces*

Cutting force modeling is essential to optimize the machining operation. In many cases, the cutting force is assumed to reduce the surface roughness values of workpieces and increase the tool wear and cutting energy consumption [27–30]. Cutting force modeling is particularly important to study the cutting performance of difficult-to-machine materials such as high-volume SiCp/Al composites. Unlike conventional homogeneous materials, the different physical and mechanical properties of the aluminum matrix and SiC particles render the cutting of SiCp/Al composites difficult to realize with good precision and high machining efficiency, thereby complicating cutting force modeling [5, 31].

Initial research on cutting force modeling in machining SiCp/Al composites started in 1985 [3], and many research articles have been published. The approaches reported in these articles can be categorized into three: experimental and empirical modeling, simulation and analytical modeling, geometrical and mechanical modeling. For the experimental and empirical modeling, a number of machining experiments were conducted with different combinations of cutting parameters and tool parameters. Several responses were measured, such as cutting force, tool wear, and surface roughness. The cutting forces were analyzed using the response surface methodology and modeled for various cutting conditions using an empirical formula [9, 32–36]. This modeling approach involved considerable time and high costs. For the simulation and analytical modeling, a finite element model is developed to simulate the machining process. The cutting parameters and tool parameters in the finite element model are regarded as the input parameters. The cutting forces generated from the simulation are the output obtained from the finite element model [37–43]. Owing to the complex and different deformed mechanisms of the aluminum matrix and SiC particles, the simulation and analytical modeling approach cannot completely characterize the phenomena occurring in the cutting of SiCp/Al composites. For geometrical and mechanical modeling, the geometrical relation between the tools and workpieces are incorporated into the mechanical model of the machining process. Cutting forces are derived from geometrical and mechanistic modeling [44, 45]. This approach can exactly predict the cutting force in advance to optimize the cutting parameter and improve the machining performance [46].

However, the study of cutting force in machining SiCp/Al composites using the geometrical and mechanical modeling approach is limited. Pramanik et al. [47] decomposed the cutting force into three components: chip formation force, plowing force, and particle fracture force. Merchant's analysis was used to calculate the chip formation force. The plowing force was calculated using the slip line field theory of plasticity. The particle fracture force was formulated using the Griffith theory. However, they did not analyze the friction force occurring on the chip-tool interface owing to the reinforcement particles. Dabade et al. [48] presented a mechanical model based on a novel chip-tool interface friction in machining SiCp/Al composites. In this model, the friction between the chip and tool can be considered to include two-body abrasion and three-body rolling induced by SiC particles. However, they did not consider the plowing force generated from the aluminum matrix owing to the tool nose. Sikder and Kishawy presented a cutting force model in machining SiCp/Al composites. The shear force and plowing force generated from the aluminum matrix, particle fracture generated from the reinforcement particles, and friction force occurring on the chip-tool interface were considered to estimate the cutting forces [49]. However, the authors did not consider the effect of the cutting edge radius on shear angle, which can reduce the accuracy of assessing deformation.

2.1.3 The Study of Material Removal Rate

Beside surface roughness, *MMR* is of greatest importance to improve product performance. As known, high enough *MRR* can decrease the time and energy consumption for removing material, which results in product cost decrement. But, the machining parameters for high *MMR* usually are contradictory with those for low surface roughness values [50]. It is necessary to optimize machining parameters considering better surface roughness and *MMR* synthetically [51, 52]. Little literatures reported the machining parameters optimization for the favorable surface roughness and *MMR* on machining composites. Surinder conducted an orthogonal array of turning unidirectional glass fiber reinforced plastics composite considering multiple performances with surface roughness and *MMR*. It was found that depth of cut followed by cutting speed was the most influential factor [53]. Prakash conducted a L27 orthogonal array of turning Aluminium/Rock dust composites. Weight fraction, size of reinforced particles, cutting speed, depth of cut, and feed were chosen as input parameters, surface roughness and *MMR* were chosen as outputs [54]. In the above literatures Taguchi and Grey Relational Analysis both were used to calculated the relational coefficient between surface roughness and *MMR*. The optimal machining parameters combination was different from those only considering single performance.

2.1.4 The Study of High Volume Fraction SiCp/Al Composites

Mechanical properties vary greatly with respect to variations in the volume fraction of SiC particles, including the hardness, bending strength, and fracture toughness, especially when the volume fraction is greater than 30 vol.% [55, 56]. Limited by the relatively new preparation technology of mixing SiC particles into aluminium matrices, research on the machining of high-volume-fraction SiCp/Al composites has appeared only in the last five years.

Huang et al. reported the machinability of drilling high-volume fractions SiCp/Al composites (> 40 vol.% particles) for the first time [15]. In his research, feed rate was found to be the main influencing parameters for drilling forces, tool wear, and surface roughness in drilling 55–57% SiCp/Al composites compared with cutting speed. Huang et al. also compared the tool wear morphology, wear resistance, cutting force and surface roughness in high-speed milling SiCp/Al composites with different volume fractions. The experiment results shown that the machinability of 56% SiCp/Al composites is worse than the other three composites at the same cutting parameters [57]. Wang et al. [58] investigated the influence on surface roughness, surface residual stress and 3D morphology of milling speed, feed rate and depth of cut in milling 65% SiCp/Al composite. It was revealed that milling speed and axial depth of cut are respectively the significant milling parameters for the corresponding surface roughness and residual stress. For the some composites, Wang et al. [59] reported the tool wear increased dramatically with the increasing milling speed, and the surface roughness is vast different for different milling speed. Yang et al. first discussed the impact of diamond grain sizes on surface roughness, surface topography and tool wear in milling high-volume fraction SiCp/Al composites with 65%

volume fraction and 80–120 μm grain size of SiC particles [60]. Wang et al. modeled cutting force generated from three zones for turning 45% SiCp/Al composites. The friction force was calculated considered sliding friction force and rolling friction force [61]. Duan et al. studied five different cooling and lubrication approaches of dry, liquid nitrogen (LN2), minimum quantity lubrication (MQL), cutting oil (Oil) and emulsion environments to turning 50% SiCp/Al composites. The experimental results shown MQL could reduce the boundary wear, tool breakage, and major flank wear and was the suitable cooling and lubrication mode in turning high-volume SiCp/Al composites [62].

The available literature clearly shows that limited efforts have been made to investigate the machinability of high-volume-fraction SiCp/Al composites. In particular, little work has been done to study surface roughness in machining high-volume-fraction SiCp/Al composites to date. Nevertheless, surface roughness is usually uncontrollable when turning high-volume-fraction SiCp/Al composites due to the complicated deformation mechanism and mechanical properties, which results in an increase in the rate of nonconforming products. It is necessary to investigate the parameters affecting surface roughness to optimise the cutting parameter selection. Therefore, high volume fraction SiCp/Al composite material is taken as the research object. Surface roughness, material removal rate and cutting force were studied. Taguchi method and analysis of variance (ANOVA) were used to explain the process of intelligent optimization of cutting parameters.

2.2 Sensitivity Analysis and Cutting Parameter Optimization Based on Taguchi Method

2.2.1 Design of Experiments

For this research, an orthogonal experiment array $L_{16}\ (4)^4$ was designed to conduct turning experiments. The four cutting parameters of the (*A*) nose radius, (*B*) cutting depth, (*C*) feed rate, and (*D*) cutting speed were the input factors, and R_a was the output response. Four levels of each cutting parameter were selected for the turning experiments based on the literature [11, 61]. Accordingly, the selected level values of the cutting parameters and the orthogonal array $L_{16}\ (4)^4$ are given in Tables 2.1 and 2.2 respectively.

For surface roughness, a smaller value is better. The S/N ratio η is obtained using Eq. 2.1.

$$\eta = -10\log_{10}\left(\frac{1}{\mathrm{n}}\sum_{i=1}^{n} R_i^2\right) \tag{2.1}$$

where $i = 1, 2, 3, \ldots, n$ (here, $n = 5$) and R_i is the $\boldsymbol{R_a}$ value.

Table 2.1 Parameters and their levels

Parameter	Notation	Levels			
		1	2	3	4
r_ε (mm)	*A*	0.2	0.4	0.6	0.8
a_p (mm)	*B*	0.1	0.15	0.2	0.25
f (mm/rev)	*C*	0.02	0.05	0.08	0.12
v_c (m/min)	*D*	150	200	250	300

Table 2.2 Orthogonal array L16 $(4)^4$ [63]

No	*A*	*B*	*C*	*D*
1	1	1	1	1
2	1	2	2	2
3	1	3	3	3
4	1	4	4	4
5	2	1	2	3
6	2	2	1	4
7	2	3	4	1
8	2	4	3	2
9	3	1	3	4
10	3	2	4	3
11	3	3	1	2
12	3	4	2	1
13	4	1	4	2
14	4	2	3	1
15	4	3	2	4
16	4	4	1	3

2.2.2 Experimental Results and Discussion

Several measurements were taken at every location. The average surface roughness is calculated and noted as R_1-R_5 after removing the abnormal data, which are listed in Table 2.3. The average R_a, noted as $\overline{R}$, is listed in Table 2.3. The S/N ratio was calculated using Eq. 2.1 and appeared in Table 2.3 as the final column.

Because a larger S/N ratio means a smaller R_a, the optimum cutting conditions among the 16 experiments can be obtained. From Table 2.3, the largest S/N ratio is 11.28 dB, and the corresponding optimum cutting conditions are $r_\varepsilon = 0.6$ mm, $a_p = 0.2$ mm, $f = 0.02$ mm/r, and $v_c = 200$ m/min ($A_3B_3C_1D_2$).

The mean S/N ratios of all parameters on each level are listed in Table 2.4. The mean S/N ratio for the 16 experiments is 4.156 dB.

Table 2.3 Surface roughness and S/N ratio

No	Surface roughness Ra (μm)						S/N ratio η_j (dB)
	R_1	R_2	R_3	R_4	R_5	$\overline{R}$	
1	0.273	0.27	0.28	0.285	0.282	0.278	11.12
2	0.669	0.657	0.644	0.679	0.661	0.662	3.58
3	1.467	1.416	1.467	1.483	1.447	1.456	– 3.26
4	2.292	2.255	2.309	2.302	2.282	2.288	– 7.19
5	0.383	0.381	0.416	0.447	0.423	0.410	7.73
6	0.38	0.386	0.347	0.371	0.354	0.368	8.69
7	1.133	1.19	1.145	1.12	1.137	1.145	– 1.18
8	0.756	0.738	0.758	0.744	0.774	0.754	2.45
9	0.603	0.616	0.599	0.635	0.622	0.615	4.22
10	0.816	0.833	0.832	0.833	0.826	0.828	1.64
11	0.272	0.269	0.277	0.271	0.276	0.273	11.28
12	0.408	0.425	0.416	0.414	0.377	0.408	7.78
13	0.852	0.873	0.837	0.873	0.865	0.860	1.31
14	0.671	0.618	0.678	0.634	0.634	0.647	3.78
15	0.509	0.543	0.516	0.532	0.525	0.525	5.59
16	0.369	0.359	0.348	0.354	0.35	0.356	8.97

Table 2.4 Mean S/N ratio

Notation	Parameter	Mean S/N ratio (dB)				
		1	2	3	4	Max.–Min
A	r_ε	1.061	4.422	6.229	4.912	5.168
B	a_p	6.094	4.42	3.107	3.003	3.091
C	f	10.012	6.171	1.796	– 1.355	11.367
D	v_c	5.374	4.654	3.768	2.828	2.546
Total mean S/N ratio (dB) = 4.156						

From Table 2.4, f is the factor that has the highest variability of 11.367. According to the Taguchi approach, the factor with the largest variability in the S/N ratio affects R_a significantly. Thus, increasing the feed rate significantly increases the R_a values.

Figure 2.1 presents the effects of the four parameters on the surface roughness and the S/N ratio. Figure 2.1 shows that the increase in feed rate increases the surface roughness under all investigated conditions. Additionally, the increase in cutting speed and cutting depth causes greater roughness values under all investigated conditions. As previously mentioned, the cutting speed and cutting depth have less effect than the feed rate on the surface roughness. However, the rate of increase in surface roughness at low cutting speeds is less than that at high speeds. This phenomenon was attributed to the high temperature in the cutting zone. A higher cutting speed increases the temperature, which causes a decrease in the bonding effect between the SiC particles and the A_l matrix. The effect of the increase in nose radius varies across the cutting conditions. The increase in nose radius decreases the surface roughness at a small nose radius. When the nose radius is greater than 0.6 mm, the increase in the nose radius increases the surface roughness.

ANOVA was carried out on the S/N ratio to identify the significant cutting parameter and its contribution to surface roughness. The results of ANOVA are shown in Table 2.5 (where F is the variance ratio).

From Table 2.5, the F values of r_ε, a_p and f are 38.624, 16.641 and 197.651, respectively, which are greater than the F distribution table value (F1,3 = 10.13 at the 5% level of significance). Accordingly, r_ε, a_p and f are confirmed as significant parameters. The F value of v_c is 9.699. The effect of the cutting parameters is ordered as $f > r_\varepsilon > a_p > v_c$, which is different from the results reported in other research [26]. Baburaj et al. found that the influence order of the cutting parameters on R_a is $v_c > f > a_p > r_\varepsilon$; in particular, r_ε was considered to have no effect and to be an error factor.

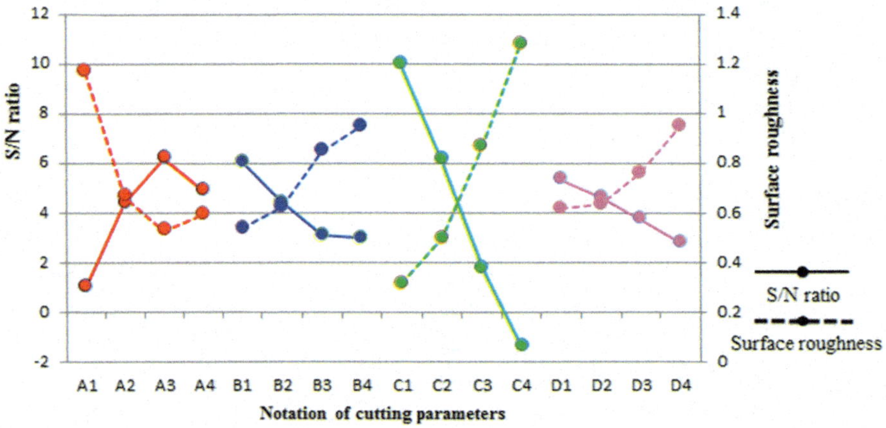

Fig. 2.1 Graph of surface roughness and S/N ratio

Table 2.5 ANOVA for surface roughness

Source	Sum of squares	Mean square	F value	Contribution (%)
r_ε	58.07	19.356	38.624	14.71
a_p	25.02	8.339	16.641	6.34
f	297.15	99.051	197.651	75.26
v_c	14.58	4.860	9.699	3.69
Error	6.01	0.501		
Total	400.83			

Table 2.6 Results comparison

Material	Fabrication method	Particle size (μm)	Nose radius	Depth of cut	Feed rate	Cutting speed	Influence sequence
All–5% SiCpl–5% fly ash [26]	Stir casting	150	■	▲	▼	▲→▼→▲	$v_c > f > a_p > r$
All–10% SiCp [19]	Stir casting	20–40	●	▲	▲	▼	●
All–20% SiCp [20]	Stir casting	25	●	▲	▲	▼	●
All–30% SiCp [11]	●	3	●	▲	▲	▼	●
Al–45% SiCp	●	5	▼→▲	▲	▲	▲	$f > r_\varepsilon > a_p > v_c$

▲: Surface roughness increases with increasing cutting parameters
▼: Surface roughness decreases with increasing cutting parameters
■: Approximately unchanged
●: No mentioned

Generally, the influence of cutting parameters on surface roughness varies greatly with respect to changes in the volume fraction of SiC particles. Table 2.6 compares the results from machining four different volume fraction SiCp/Al composites.

2.2.3 *Predicting the Optimum Surface Roughness*

From Fig. 2.1, the combination of $A_3B_1C_1D_1$ is the optimum level combination, from which the largest S/N ratio will be obtained. Accordingly, the smallest surface roughness can be obtained under the cutting conditions when r_ε is 0.6 mm, a_p is 0.1 mm, f is 0.02 mm/rev, and v_c is 150 m/min. The largest S/N ratio and the corresponding surface roughness can be calculated using Eqs. 2.2 and 2.3, respectively.

$$S_e = \overline{T_A^3} + \overline{T_B^1} + \overline{T_C^1} + \overline{T_D^1} - 3 * \overline{S} = 15.241\ \text{dB} \tag{2.2}$$

$$Ra_e = \sqrt{10^{[S_e/(-10)]}} = 0.173\ \mu\text{m} \tag{2.3}$$

where $\overline{T_A^3}$ is the mean S/N ratio for notation A on level 3, $\overline{T_B^1}$ is the mean S/N ratio for notation B on level 1 , $\overline{T_C^1}$ is the mean S/N ratio for notation C on level 1, $\overline{T_D^1}$ is the mean S/N ratio for notation D on level 1, and $\overline{S}$ is the total mean S/N ratio for the 16 experiments.

By using the polynomial regression equation, the surface roughness can be predicted as follows:

$$\begin{aligned} Ra &= 0.3073 - 0.4454r_{\varepsilon} + 0.2728a_{\mathrm{p}} + 0.0848v_{\mathrm{c}} \\ &+ 0.1795r_{\varepsilon}^2 - 0.0077a_{\mathrm{p}}^2 + 0.0791f^2 + 0.024v_{\mathrm{c}}^2 \\ &- 0.076r_{\varepsilon}a_{\mathrm{p}} - 0.1226r_{\varepsilon}f - 0.0432r_{\varepsilon}v_{\mathrm{c}} + 0.0364a_{\mathrm{p}}f \\ &- 0.0434a_{\mathrm{p}}v_{\mathrm{c}} + 0.0236fv_{\mathrm{c}}^2 \end{aligned} \tag{2.4}$$

A comparison of the predicted surface roughness results and the experimental results is shown in Table 2.7. It is observed that the predicted value is very close to the experimental value. The maximum predicted deviation is – 10.00% under the conditions when r_{ε} is 0.4 mm, a_{p} is 0.1 mm, f is 0.05 mm/rev, and v_{c} is 250 m/min. The minimum predicted deviation is 0.38% under the conditions when r_{ε} is 0.2 mm, a_{p} is 0.25 mm, f is 0.12 mm/rev, and v_{c} is 300 m/min. The average predicted deviation is 4.99%, which shows that the predicted value agrees well with the experimental value.

We performed two experiments to check the effectiveness and accuracy of the Taguchi approach and the polynomial regression equation. The predicted optimum R_{a} values using the Taguchi approach and polynomial regression equation are given in Table 2.8.

Table 2.7 Experimental results and predicted results of surface roughness

No	Experiment result	Predicted result	Deviation (%)
1	0.278	0.2692	– 3.17
2	0.662	0.721	8.91
3	1.456	1.3968	– 4.07
4	2.288	2.2966	0.38
5	0.41	0.369	– 10.00
6	0.368	0.3768	2.39
7	1.145	1.1358	– 0.80
8	0.754	0.7948	5.41
9	0.615	0.6558	6.63
10	0.828	0.8354	0.89
11	0.273	0.2654	– 2.78
12	0.408	0.3674	– 9.95
13	0.86	0.8304	– 3.44
14	0.647	0.6836	5.66
15	0.525	0.4888	– 6.90
16	0.356	0.386	8.43
Average deviation			4.99

Table 2.8 Optimum cutting parameters and surface roughness

No	Predicted approach	Notation	Leve	Cutting parameters		Optimum surface roughness (μm)		Deviation (%)
				Parameter	Value	Predicted	Experiment	
1	Taguchi approach	*A*	3	r_ε	0.6 mm	0.173	0.192	− 9.90
		B	1	a_p	0.1 mm			
		C	1	f	0.02 mm/r			
		D	1	v_c	150 m/min			
2	Polynomial regression equation	*A*	3	r_ε	0.6 mm	0.105	0.113	− 7.08
		B	4	a_p	0.25 mm			
		C	1	f	0.02 mm/r			
		D	4	v_c	300 m/min			

The experimental results show that the predicted deviation in both approaches is less than 10%. From ANOVA, r_ε and f provide an 89.97% contribution to R_a. The optimum settings for r_ε (0.6 mm) and f (0.02 mm/rev) are noted to be the same in the Taguchi approach and in the polynomial regression equation, which is consistent with the results of ANOVA. With respect to a_p and v_c, the optimum setting in the Taguchi approach is the minimal value, whereas this is the maximum value in the polynomial regression equation. It is expected that the increase in a_p and v_c could decrease the R_a values under the conditions when r_ε is 0.6 mm and f is 0.02 mm/rev. Table 2.8 verifies that the polynomial regression equation gives better predicted results than the Taguchi approach for the optimum surface roughness.

2.2.4 Conclusion

The surface roughness of machined high-volume-fraction SiCp/Al composites was investigated through the Taguchi approach. L_{16} $(4)^4$ orthogonal experiments were designed and conducted, covering the four cutting parameters of r_ε, a_p, f and v_c, each at four levels, to investigate the effect on surface roughness. The following conclusions can be drawn from the results.

(1) The parameters f, r_ε and a_p were significant factors for the surface roughness of machined high-volume-fraction SiCp/Al composites.
(2) From ANOVA, the contribution of f and r_ε to the surface roughness was 89.97%, whereas the contribution of a_p and v_c was only 10.03%.
(3) The mathematical model for predicting the surface roughness, which was developed based on a polynomial regression equation, was shown to be valid and accurate.

Table 2.9 Cutting parameters for turning and their levels

Symbol	Parameters	Levels			
		1	2	3	4
A	r_ε (mm)	0.2	0.4	0.6	0.8
B	a_p (mm)	0.1	0.15	0.2	0.25
C	f (mm/r)	0.02	0.05	0.08	0.12
D	v_c (m/min)	150	200	250	300

(4) A confirmation experiment showed that the polynomial regression equation performed better than the Taguchi approach in predicting the optimum surface roughness.

2.3 Multi-objective Cutting Parameter Optimization Based on Grey Correlation Method

2.3.1 Experimental Design

Taguchi's orthogonal experiment design is used in this work [64]. The design involves changes in r_ε, a_p, f and v_c at four different levels, as shown in Table 2.9.

In total 16 experiments were conducted. The output responses were cutting force (*F*), surface roughness (*Ra*) and *MRR*. For surface roughness, every measuring operation was performed on five different locations, which is shown in Fig. 2.2. Several measurements for each location were taken to calculate the average surface roughness noted as R_1—R_5 after removing the abnormal data. The average cutting force can be obtained on the stable stage of measured cutting force, which is the resultant force of main cutting force F_z, feed force F_y and radial force F_x shown in Fig. 2.3. The value of *MRR* can be calculated by Eq. 2.5 [65]. Experimental results on R_a, *F* and *MRR* are listed in Table 2.3 (Table 2.10).

$$MRR = \frac{1000 v_c \times a_p \times f}{60} \tag{2.5}$$

2.3.2 Results and Discussion

S/N ratios are usually used to evaluate the effect of various parameters on the output quality as an important indicator. As the objective of this study is to minimize surface roughness, cutting force, to maximize *MMR*, the corresponding S/N ratio can be calculated by using Eqs. 2.6–2.8 [66], which are shown in Table 2.11.

Fig. 2.2 Five measuring locations

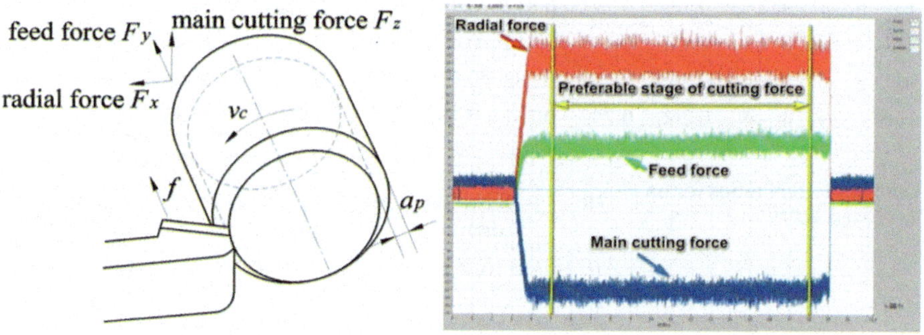

Fig. 2.3 Preferable stage of cutting force

$$S/N = -10\lg\left(\frac{1}{n}\sum_{i=1}^{n} Ra_i^2\right) \tag{2.6}$$

$$S/N = -10\lg\left(\frac{1}{m}\sum_{i=1}^{m} F_i^2\right) \tag{2.7}$$

$$S/N = -10\lg\left(\frac{1}{m}\sum_{i=1}^{m} \frac{1}{MRR_i^2}\right) \tag{2.8}$$

where n, m is the number of measurements ($n = 5$ and $m = 1$).

Effect of cutting parameters on surface roughness and its S/N ratio are shown in Fig. 2.4. Figure 2.4 shows that increasing depth of cut and feed rate would increase the values of surface roughness, which are in fine agreement with that for machining

Table 2.10 Experimental results on R_a, F and MRR

No	A	B	C	D	Measured surface roughness					Average R_a (μm)	F (N)	MRR (mm³/s)
					R_1	R_2	R_3	R_4	R_5			
1	1	1	1	1	0.308	0.312	0.301	0.3	0.296	0.303	7.76	5.0
2	1	2	2	2	0.727	0.719	0.759	0.742	0.734	0.736	14.11	25.0
3	1	3	3	3	1.786	1.697	1.746	1.739	1.726	1.739	23.18	66.7
4	1	4	4	4	2.933	2.893	2.919	2.899	2.921	2.913	36.76	150.0
5	2	1	2	3	0.336	0.341	0.312	0.337	0.326	0.330	14.31	20.8
6	2	2	1	4	0.251	0.236	0.254	0.24	0.258	0.248	13.53	15.0
7	2	3	4	1	1.378	1.338	1.352	1.346	1.37	1.357	30.92	60.0
8	2	4	3	2	0.497	0.476	0.514	0.469	0.488	0.489	30.56	66.7
9	3	1	3	4	0.588	0.602	0.542	0.558	0.601	0.578	17.73	40.0
10	3	2	4	3	0.808	0.775	0.781	0.809	0.757	0.786	28.42	75.0
11	3	3	1	2	0.329	0.329	0.311	0.333	0.324	0.325	15.93	13.3
12	3	4	2	1	0.423	0.418	0.416	0.413	0.432	0.420	25.57	31.3
13	4	1	4	2	0.762	0.791	0.812	0.756	0.749	0.774	24.51	40.0
14	4	2	3	1	0.445	0.437	0.417	0.412	0.402	0.423	26.49	30.0
15	4	3	2	4	0.268	0.269	0.288	0.288	0.284	0.279	25.35	50.0
16	4	4	1	3	0.266	0.252	0.261	0.263	0.258	0.260	22.85	20.8

Table 2.11 S/N ratios for on R_a, F and MRR

No	S/N ratio for R_a (dB)	S/N ratio for F (dB)	S/N ratio for MRR (dB)
1	10.37	− 17.8	13.98
2	2.66	− 23.0	27.96
3	− 4.81	− 27.3	36.48
4	− 9.29	− 31.3	43.52
5	9.12	− 23.1	26.36
6	11.84	− 22.6	23.52
7	− 2.65	− 29.8	35.56
8	6.21	− 29.7	36.48
9	4.73	− 25.0	32.04
10	1.87	− 29.1	37.50
11	8.80	− 24.0	22.48
12	7.54	− 28.2	29.91
13	2.23	− 27.8	32.04
14	6.43	− 28.5	29.54
15	11.09	− 28.1	33.98
16	11.70	− 27.2	26.36
Mean	4.87	− 26.41	30.48

low-volume SiCp/Al composites [11, 19, 20], in which the volume fractions of SiC particles are 10%, 20% and 30% respectively. However, the influence of increasing depth of cut on surface roughness gradually weakens. Again the values of surface roughness would increase while cutting speed changes from 200 m/min to 300 m/min, which is different from the results reported in the above three literatures. It was proved that surface finish would be worse in turning 45% SiCp/Al composites with the higher cutting speed [24]. So, it can be concluded that a large amount of SiC particles and high cutting temperature would deteriorate surface finish in machining high-volume SiCp/Al composites compared with low-volume. Figure 2.4 shows that increasing nose radius would decrease surface roughness. Especially, while the tool nose radius is close to the cutting depth, surface roughness increases significantly [26]. It is suggested that nose radius much bigger than depth of cut should be selected for finished high-volume SiCp/Al composites. From S/N ratios, A_4, B_1, C_1 and D_1 are identified as the optimal cutting parameters combination. Generally, analysis results of surface roughness and its S/N ratio show that the selected strategy of cutting parameters for machining high-volume SiCp/Al composites are different from those for machining low-volume composites.

Effect of cutting parameters on cutting force and S/N ratio is shown in Fig. 2.5. Figure 2.5 shows that increasing nose radius, depth of cut and feed rate would increase the values of cutting force. The effect is in fine agreement with that of result from some literatures, in which cutting force was investigated in machining two types of composites with 20%, 30% SiC particles [67]. It was well stated that the increase of radial force caused by the increase of tool nose radius contributed to the increase of resultant force. In fact, it was reported that increasing cutting speed caused the decrease of cutting force from 24 m/min to 150 m/min in turning SiCp/Al composites with 20 and 8% SiC particles [5, 34]. It was proved that thermal softening of the workpiece caused the decrement of cutting force [68]. From Fig. 2.5, it is observed that the values of cutting force would increase while cutting speed increase from 200 m/min to 300 m/min, which is different from those of machining low-volume composites.

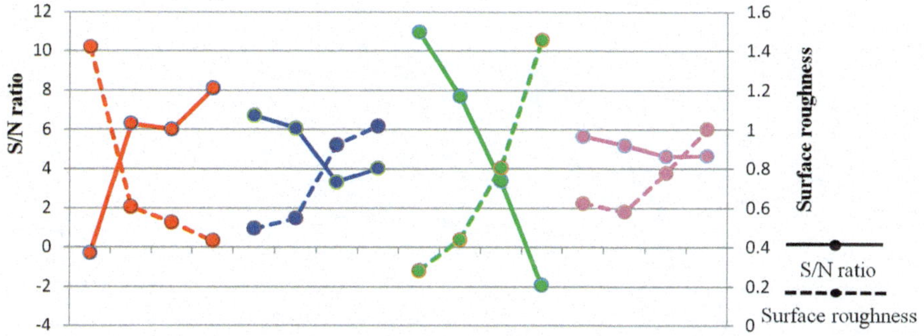

Fig. 2.4 Effect of cutting parameters on surface roughness and its S/N ratio

Therefore it can be concluded that a large amount of SiC particles weakened thermal softening of workpiece and caused the increase of cutting force. From S/N ratios, A_1, B_1, C_1 and D_1 are identified as the optimal cutting parameters combination, which is different from those for the minimal surface roughness.

Figure 2.6 confers for a clear view on the effect of cutting parameters on *MRR*. Further the optimal parameters are listed as B_4, C_4 and D_4 for the maximal *MRR*.

The behavior of R_a and cutting force, in respect of each other, is depicted in Fig. 2.7. The surface roughness appears to be lower at smaller cutting force which is justifiable by vibration texture of finished surface. More SiC particles were fractured to be irregular rolling-body on the interface between the vice flank face and finished face due to the bigger force, which could increase the fluctuation of cutting force.

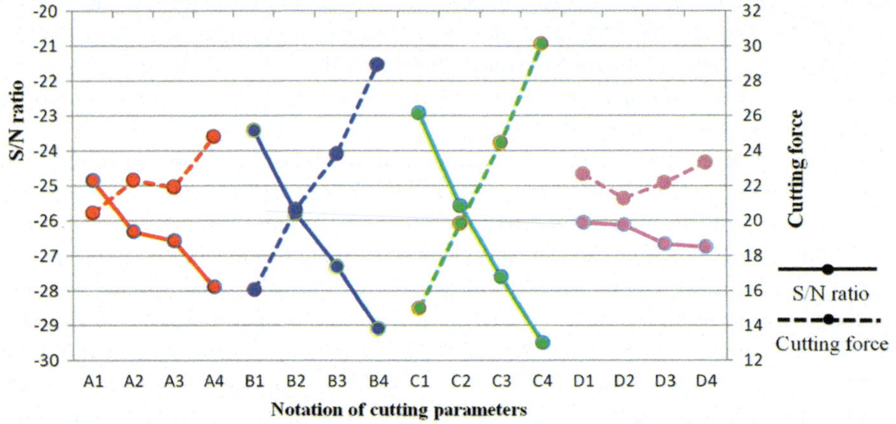

Fig. 2.5 Effect of cutting parameters on cutting force and its S/N ratio

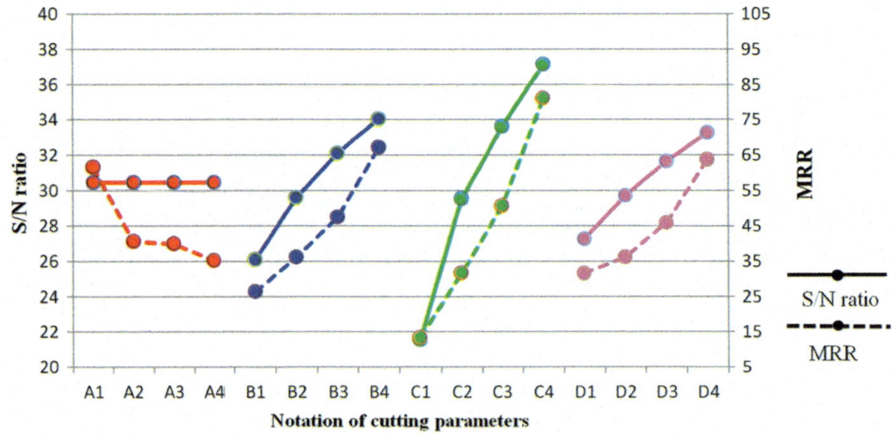

Fig. 2.6 Effect of cutting parameters on *MRR* and its S/N ratio

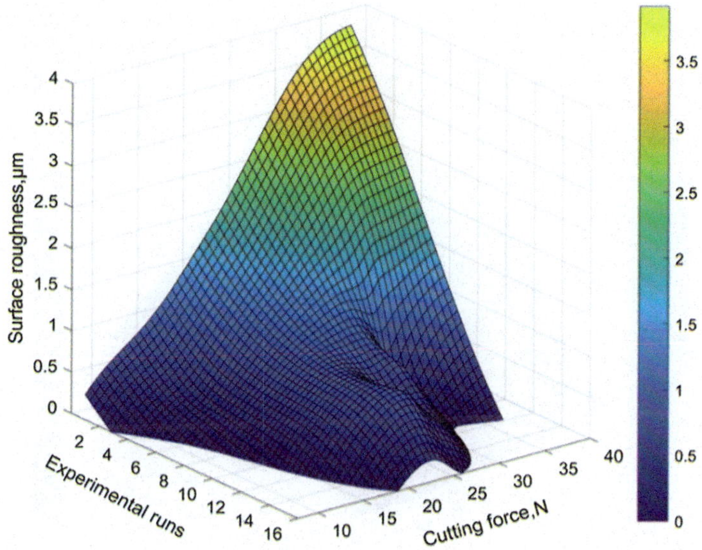

Fig. 2.7 Correlation of surface roughness and cutting force

From Fig. 2.7, it is affirmed that a low surface roughness exists below the cutting force of about 15N, and it is attributable to the smaller fluctuation of cutting force. The relationship between surface roughness and *MRR* is shown in Fig. 2.8. The surface finish appears to be worse at higher *MRR*. From Fig. 2.8, it is confirmed again that higher *MRR* and lower surface roughness are usually contradict each other.

In summary, the optimal parameters combination for the minimal surface roughness is $A_4B_1C_1D_1$, is $A_1B_1C_1D_1$ for the minimal cutting force, is $A_1B_4C_4D_4$ for the maximal *MRR*, which is completely different each other. It is necessary to optimize the cutting parameters considering the trade-off between surface roughness, cutting forces and *MRR*.

2.3.3 Multi-objective Optimization

The current study intends to find the optimal parameters combination to minimize surface roughness and cutting force, maximize *MMR*. Accordingly, the proposed optimized mathematical model can be stated as:

$$\begin{cases} Min\ Ra(r_\varepsilon, a_p, f, v_c) \\ Min\ F(r_\varepsilon, a_p, f, v_c) \\ Max\ MRR(r_\varepsilon, a_p, f, v_c) \end{cases} \tag{2.9}$$

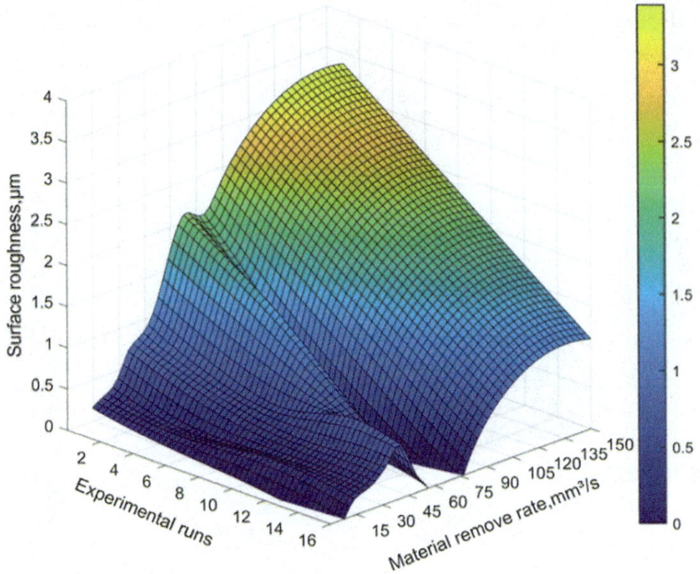

Fig. 2.8 Correlation of surface roughness and *MRR*

To facilitate to optimize cutting parameters considering the trade-off between surface roughness, cutting forces and MRR, multi-objective optimization can be converted to be flexible objective optimization, which can be presented by the following [69]:

$$\min MO(r_\varepsilon, a_p, f, v_c) = \omega_1 R_a + \omega_2 F + \omega_3 MRR \tag{2.10}$$

Subject to the following constraints:

$$r_\varepsilon \in \{0.2, 0.4, 0.6, 0.8\} \tag{2.11}$$

$$0.1 \le a_p \le 0.25 \tag{2.12}$$

$$0.02 \le \text{f} \le 0.12 \tag{2.13}$$

$$150 \le v_c \le 300 \tag{2.14}$$

ω_1, ω_2 and ω_3 are the corresponding weight factors of surface roughness, cutting force, and *MRR*, which can be obtained with respected to optimization condition. Non-uniform dimensions of surface roughness, cutting force, and *MRR* make multi-objective optimization complex. Grey relational analysis (GRA) is an ideal method to solve complex interrelationships among multiple responses. Also, GRA was proved to be an efficient approach for cutting parameters optimization [70, 71]. There are some steps should be followed when applying GRA [72].

(1) The experimental data of R_a, F and MRR are normalized to prevent the influence of using different elements to reduce variability. Equation 2.15 is used to normalize the original sequence for 'higher the better' with respect to MRR. Equation 2.16 is used to normalize the original sequence for 'lower the better' with respect to surface roughness and cutting force.

$$Z_i(k) = \frac{y_i(k) - \min(y_i(k))}{\max(y_i(k)) - \min(y_i(k))} \tag{2.15}$$

$$Z_i(k) = \frac{\max(y_i(k)) - y_i(k)}{\max(y_i(k)) - \min(y_i(k))} \tag{2.16}$$

(2) Calculate the grey relational coefficient (GRC) of normalized value.

$$\xi_i(k) = \frac{\Delta\min + \xi\Delta\max}{\Delta_{oi}(k) + \xi\Delta\max} \tag{2.17}$$

$i = 1, 2\ldots n$, n is the number of experiments and n equals 16; $k = 1, 2\ldots m$, m is the number of responses and m equals 3; $Z_o(k)$ is the reference sequence ($Z_o(k) = 1$); $\Delta_{oj}(k)$ is the deviation series of the reference series $Z_i(k)$, i.e. $\Delta_{oi}(k) = \|Z_o(k) - Z_i(k)\|$; $\Delta\max = \max(\Delta_{oi}(k))$ is the maximum value of $\Delta_{oi}(k)$; $\Delta\min = \min(\Delta_{oi}(k))$ is the minimum value of $\Delta_{oi}(k)$; ξ is the distinguishing coefficient which is defined in the range of $0 \le \xi \le 1$ and the value of ξ equals 0.5 normally.

(3) Calculate the grey relational grade (GRG) by using the normalized weight factor.

$$\gamma_i = \sum_{k=1}^{m} w(k)\xi_i(k) \tag{2.18}$$

$w(k)$ represents weight factor for selected response and $\sum_{k=1}^{m} w(k) = 1$.

The weight factor can be calculated by using the following equations:

$$R_i(k) = \max\left(\frac{\sum_{j=1}^{o}\xi_{ij}(k)}{o}\right) - \min\left(\frac{\sum_{j=1}^{o}\xi_{ij}(k)}{o}\right) \tag{2.19}$$

$$w(k) = \frac{\sum_{i=1}^{p} R_i(k)}{\sum_{k=1}^{m}\sum_{i=1}^{p} R_i(k)} \tag{2.20}$$

$\xi_{ij}(k)$ is the average GRC of cutting parameter i at level j on response k; $i = 1, 2\ldots p$, p is the number of cutting parameters and p equals 4; $j = 1, 2\ldots o$, o is the number of levels of each parameters and o equals 4; $R_i(k)$ is the range of GRC of cutting parameter i on response k; m is number of selected responses, $m \in \{2, 3\}$;

Three different conditions are suggested to optimize cutting parameters listed as follows:

Condition 1: Weighting optimization considering R_a and F ($\omega_1 = w(1)$, $\omega_2 = w(2)$, $\omega_3 = 0$).

Condition 2: Weighting optimization considering R_a and MRR ($\omega_1 = w(1)$, $\omega_3 = w(3)$, $\omega_2 = 0$).

Condition 3: Weighting optimization considering R_a, F and MRR ($\omega_1 = w(1)$, $\omega_2 = w(2)$, $\omega_3 = w(3)$).

$\xi_i(1)$, $\xi_i(2)$ and $\xi_i(3)$ in Table 2.12 mean GRC of surface roughness, cutting force and *MRR* respectively calculated by Eqs. 2.15–2.17. According to Table 2.12 and Eq. 2.18, the analysis of GRC of surface roughness, cutting force and *MRR* is shown Table 2.13. Referring to the different optimization conditions, the expressions of GRG γ_i from condition 1 to condition 3 are presented in Eqs. 2.21–2.23.

$$\gamma_i = 0.507993\xi_i(1) + 0.492007\xi_i(2) \tag{2.21}$$

$$\gamma_i = 0.532\xi_i(1) + 0.478\xi_i(3) \tag{2.22}$$

Table 2.12 Calculated results of GRC $\xi_i(k)$ and GRG γ_i

No	$\xi_i(1)$	$\xi_i(2)$	$\xi_i(3)$	Condition 1		Condition 2		Condition 3	
				γ_i	Rank	γ_i	Rank	γ_i	Rank
1	0.96036	1	0.333333	0.979863	1	0.666911	6	0.780176	1
2	0.731942	0.695444	0.367089	0.713985	7	0.561191	14	0.606843	7
3	0.471932	0.484626	0.46534	0.478178	14	0.468847	16	0.474213	15
4	0.333333	0.333333	1	0.333333	16	0.645333	9	0.539239	13
5	0.942029	0.688836	0.359445	0.817456	3	0.66938	4	0.675995	3
6	1	0.715343	0.349398	0.859947	2	0.695518	2	0.702259	2
7	0.545771	0.385024	0.446154	0.466682	15	0.49915	15	0.460342	16
8	0.846838	0.38874	0.46534	0.621451	11	0.668297	5	0.573235	11
9	0.801504	0.592562	0.39726	0.698703	8	0.612318	11	0.6056	8
10	0.712376	0.4124	0.491525	0.564786	13	0.609018	12	0.542159	12
11	0.945371	0.639612	0.346558	0.794935	4	0.665127	7	0.65645	4
12	0.885676	0.448777	0.379184	0.670719	9	0.648638	8	0.580676	9
13	0.716976	0.464	0.39726	0.59251	12	0.567349	13	0.532205	14
14	0.883914	0.436353	0.376623	0.663711	10	0.646502	10	0.575041	10
15	0.977264	0.451854	0.42029	0.718759	6	0.7166	1	0.626574	5
16	0.991075	0.49003	0.359445	0.744557	5	0.695472	3	0.625612	6

Table 2.13 GRC analysis of each response

	Notation	Nose radius	Depth of cut	Feed rate	Cutting sped
Surface roughness	1	0.6243920	0.8552170	0.9742020	0.8189300
	2	0.8336600	0.8320580	0.8842280	0.8102820
	3	0.8362320	0.7350850	0.7510470	0.7793530
	4	0.8923070	0.7642310	0.5771140	0.7780250
	$R_i(k)$	0.2679150	0.1201320	0.3970880	0.0409050
	$\sum_i^4 R_i(k)$	0.8260400			
Cutting force	1	0.628351	0.68635	0.711246	0.567539
	2	0.544486	0.564885	0.571228	0.546949
	3	0.523338	0.490279	0.47557	0.518973
	4	0.460559	0.41522	0.398689	0.523273
	$R_i(k)$	0.1677920	0.2711300	0.3125570	0.0485660
	$\sum_i^4 R_i(k)$	0.8000450			
MRR	1	0.541441	0.371825	0.347183	0.383824
	2	0.405084	0.396159	0.381502	0.394062
	3	0.403632	0.419586	0.426141	0.418939
	4	0.388405	0.550992	0.583735	0.541737
	$R_i(k)$	0.153036	0.179168	0.236551	0.157913
	$\sum_i^4 R_i(k)$	0.726668			

$$\gamma_i = 0.351095\xi_i(1) + 0.340046\xi_i(2) + 0.308859\xi_i(3) \tag{2.23}$$

The calculated GRG values and their ranks are as listed in Table 2.12 where in the higher GRG means for the more ideal quality characteristic. From Table 2.12, on condition 1 cutting parameters of experiment no 1 with the highest GRG signifies the optimal cutting parameters among all of experiments. On the other hand, on condition 2 and condition 3 the better cutting parameters combinations belong to experiment no 15 and no 1 respectively.

Table 2.13 shows the GRG of cutting parameters on four levels on the three conditions. From Table 2.14, it is obvious that $A_2B_1C_1D_1$, $A_4B_4C_1D_4$ and $A_2B_1C_1D_4$ are identified as the optimal cutting parameters combination on conditon1, condition 2 and condition 3 respectively. These parameters combination are completely different each other. In particular, only the parameters combination on condition 1 is same as that for minimal surface roughness. The rank of GRG shows feed rate is the most significant parameters to affect the different optimization objectives from condition 1 to condition 3.

Some experiments were executed to confirm the optimal parameters combination for different single-objective and multi-objective. The optimal cutting parameters combinations and result of confirmatory experiments are listed Table 2.14.

Table 2.14 Response table for GRG on different conditions

		Nose radius	Depth of cut	Feed rate	Cutting speed
Condition 1	Level 1	0.62634	0.772133[a]	0.844826[a]	0.695244[a]
	Level 2	0.691384[a]	0.700607	0.73023	0.68072
	Level 3	0.682286	0.614639	0.615511	0.651244
	Level 4	0.679884	0.592515	0.489328	0.652686
	Delta	0.065044	0.179618	0.355498	0.044
	Rank	3	2	1	4
Condition 2	Level 1	0.585571	0.62899	0.680757[a]	0.6153
	Level 2	0.633086	0.628057	0.648952	0.615491
	Level 3	0.633775	0.587431	0.598991	0.610679
	Level 4	0.656481[a]	0.664435[a]	0.580213	0.667442[a]
	Delta	0.07091	0.077004	0.100544	0.056763
	Rank	3	2	1	4
Condition 3	Level 1	0.600118	0.648494[a]	0.691124[a]	0.599059
	Level 2	0.602958[a]	0.606576	0.622522	0.592183
	Level 3	0.596221	0.554395	0.557022	0.579495
	Level 4	0.589858	0.579691	0.518486	0.618418*
	Delta	0.0131	0.094099	0.172638	0.038923
	Rank	4	2	1	3

[a] Level of cutting parameters for better turning performance

The following conclusions can be drawn by analyzing confirmatory experiment results.

1. Compared with the optimization results of *Ra*, the results of multi-objective optimization can effectively improve the cutting force and *MRR* without significantly reducing the surface quality.
2. Compared with the optimization results of *F*, the optimization results on condition 1 and condition 3 showed that the *MRR* and surface quality were improved, and the increase of *F* was not very large.
3. Although the *MRR* is very high while the maximal *MRR* is treated as the target, the value of cutting force is very large and waved, which tend to bad surface quality.

It can be seen that the result of multi-objective optimization is a compromise compared with that of single-objective optimization. For finishing high-volume SiCp/Al composites, grey relational analysis is a very suitable approach to minimize surface roughness considering minimal cutting force and maximal *MRR*.

Table 2.14 Result of confirmatory experiments

Optimization condition	Parameters combination	r_ε(mm)	a_p(mm)	f(mm/r)	v_c(m/min)	Average R_a (μm)	F (N)	MRR (mm^3/s)
Min R_a	$A_4B_1C_1D_1$	0.8	0.1	0.02	150	0.231	12.12	5
Min F (No. 1)	$A_1B_1C_1D_1$	0.2	0.1	0.02	150	0.303	7.76	6.667
Max MRR	$A_0B_4C_4D_4$	–	0.25	0.12	300	–	36.76	150
Condition 1	$A_2B_1C_1D_1$	0.4	0.1	0.02	150	0.268	10.35	5
Condition 2	$A_4B_4C_1D_4$	0.8	0.25	0.02	300	0.274	24.47	25
Condition 3	$A_2B_1C_1D_4$	0.4	0.1	0.02	300	0.285	14.28	10

2.3.4 Conclusion

This study focused the machinability of high-volume SiCp/Al composite. 16 orthogonal experiments are conducted on CNC turning lathe to investigate the effect of nose radius, depth of cut, feed rate and cutting speed on surface roughness, cutting force and *MRR*. Certain relevant conclusions primed from this study are as follows:

1. The high-volume SiCp/Al composites exhibit the different machinability from low-volume composites. In particular, the increase of cutting speed from 200 m/min to 300 m/min tends to increase surface roughness and cutting force. Feed rate is still supposed to be the significant parameters affecting surface roughness followed by nose radius, depth of cut and cutting speed.
2. For turning high-volume SiCp/Al composites, many SiC particles were fractured to be irregular rolling-body on the interface between the vice flank face and finished face, which could increase the fluctuation of cutting force. It is attributable to the increase of surface roughness with the increase of cutting force.
3. The optimal parameters combinations considering the trade-off R_{a}, F and *MRR* are different from those for single-objective optimization. The optimal parameters combinations for Min R_a and Min F, Min R_a and Max *MRR* or Min R_a, Min F and Max *MRR* are also various. Furthermore, feed rate has the greatest influence on GRG among all of cutting parameters no matter what kind of conditions. Multi-objective optimization can achieve a compromise result based on surface quality and machining efficiency.

2.4 Modeling of Cutting Force Prediction in Machining High-Volume SiCp/Al Composites

2.4.1 Modeling of Cutting Force Components

The cutting force model in this study is based on the oblique cutting force model [73, 74]. Reinforcement particles are assumed to be spherical and distributed uniformly in the aluminum matrix [31, 37, 47, 48]. The model is shown in Fig. 2.9.

In this model, the cutting force is generated from three deformed zones. Shear deformed zone I is simplified as a shear plane AB [75]. Owing to the force on the undeformed part from the tool exceeding the shear strength of the SiCp/Al composites, chip formation occurs at the AB plane. The process is similar to the normal orthogonal cutting of a homogeneous material. The chip sliding on the rake face will overcome the friction from two aspects, the friction between the aluminum matrix at the bottom of chip and rank face of tool, and the friction between the SiC reinforce particles and rank face of tool. The two frictions occur at plane BD, which is called the friction deformed zone II. Owing to material deformation and displacement by the rounded part of the cutting edge, plowing occurs at the cambered surface BC,

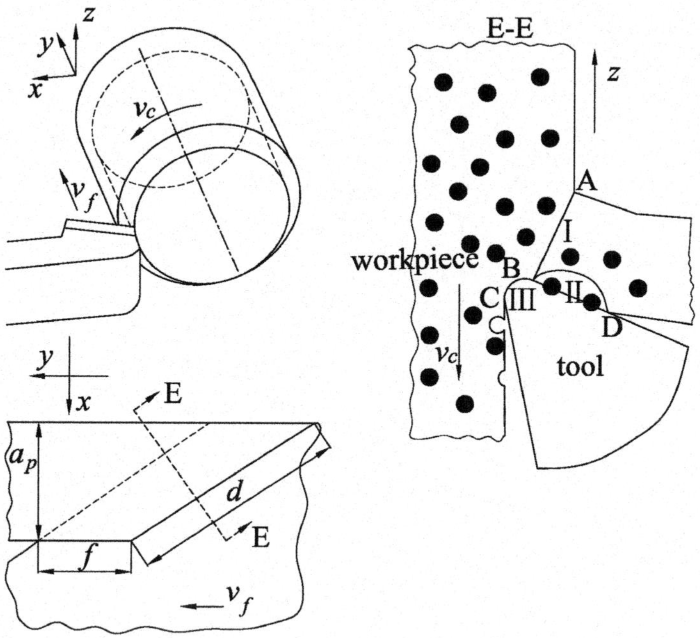

Fig. 2.9 Cutting force model of machining SiCp/Al composites

which is called plowing zone III. In this zone, only plastic deformation occurred, and no chip is formed.

(1) Analysis of cutting force components from deformed zone I.

Shear deformation occurs in zone I. Lee's analysis was selected for the present investigation to determine the chip formation forces [76]. In Lee's analysis, the cutting force on the chip is as shown in Fig. 2.10.

In Fig. 2.10, the chip is considered as a separate body in equilibrium under the action of two equal, but opposing forces $F_{\tau s}$: the force that the tool exerts on the back surface of the chip, and the force that the workpiece exerts on the base of the chip at the shear plane AB. The resultant force $F_{\tau s}$ can be given by Eq. 2.24:

$$F_{\tau s} = \sqrt{F_\tau^2 + F_c^2} = \sqrt{F_f^2 + F_n^2} \tag{2.24}$$

where F_τ is the shear force on the shear plane AB, F_c is the force that the undeformed part exerts on the shear plane AB, F_f is the sliding friction force between the back surface of the chip and tool rake face, and F_n is the normal force from the tool rake face. These forces are shown as below.

$$F_\tau = \frac{\tau_s\left[b - r_z(1 + \sin\varphi)\right]d}{\sin\varphi} \tag{2.25}$$

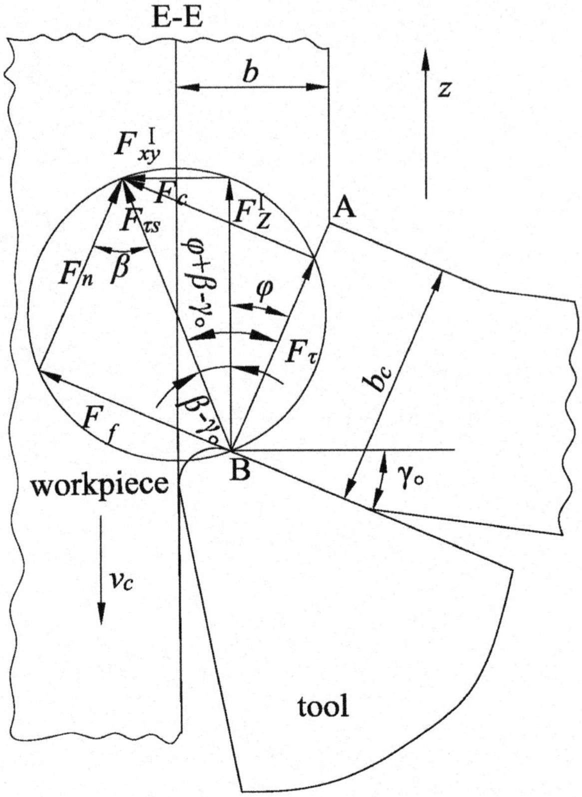

Fig. 2.10 Cutting force in the deformed zone I

$$F_c = F_\tau \tan(\varphi + \beta - \gamma_0) \tag{2.26}$$

$$F_n = \frac{F_\tau \cos\beta}{\cos(\varphi + \beta - \gamma_0)} \tag{2.27}$$

$$F_f = \frac{F_\tau \sin\beta}{\cos(\varphi + \beta - \gamma_0)} \tag{2.28}$$

where τ_s is the flow stress, b is the cut thickness, r_z is the cutting edge radius, φ is the shear angle, d is the cut width along the cut edge, β is the friction angle, and γ_0 is the rake angle. The cut thickness b and cut width along the cut edge d are given by:

$$b = f * \sin\kappa_r \tag{2.29}$$

$$d = a_p / \sin\kappa_r \tag{2.30}$$

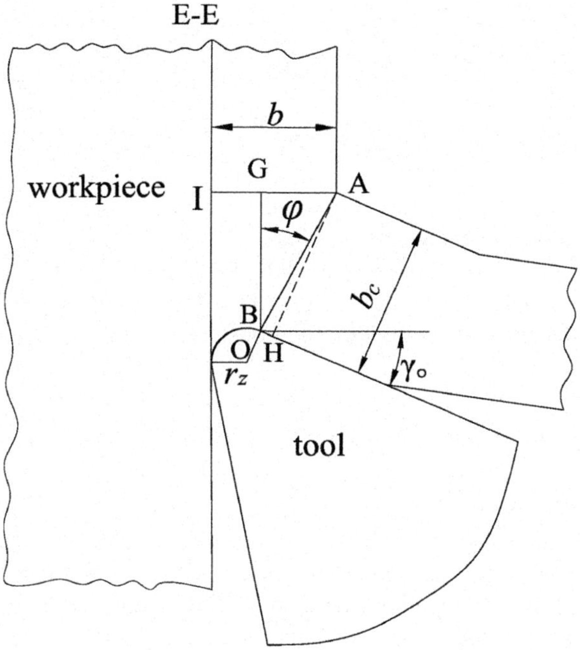

Fig. 2.11 Geometrical relation between chip thickness b_c and cutting thickness b

where f is the feed rate, a_p is the cutting depth, and k_r is the approach angle.

Chip formation results in the chip thickness b_c exceeding the cut thickness b. Considering the effect of cutting edge radius r_z, the geometrical relation between b_c and b is shown in Fig. 2.11.

From Fig. 2.11, the geometrical relation between chip thickness b_c and cutting thickness b can be expressed by

$$\frac{b_c \sin\varphi}{\cos(\varphi - \gamma_0)} + r_z(1 + \sin\varphi) = b \tag{2.31}$$

In Lee's analysis, the theoretical relation obtained for φ is

$$\varphi = \frac{\pi}{4} - \beta + \gamma_0 \tag{2.32}$$

According to Eqs. (2.24)–(2.31), $F_{\tau s}$ can be obtained, and can be decomposed into two cutting force components as follows:

$$F_{\tau s} = \sqrt{F_z^2 + F_{xy}^2} \tag{2.33}$$

Thus, the force component F_z^{I} acting on the tool in the direction of the cutting speed (Z direction in Fig. 2.10), and the force component F_{xy}^{I} in the direction perpendicular to the cut edge and cutting speed, can be determined using the equations given below:

$$F_z^{\mathrm{I}} = \frac{\tau_s\left[b - r_z(1 + \sin\varphi)\right]d\cos(\beta - \gamma_0)}{\sin\varphi\cos(\varphi + \beta - \gamma_0)} \tag{2.34}$$

$$F_{xy}^{\mathrm{I}} = \frac{\tau_s\left[b - r_z(1 + \sin\varphi)\right]d\sin(\beta - \gamma_0)}{\sin\varphi\cos(\varphi + \beta - \gamma_0)} \tag{2.35}$$

F_{xy} can be further decomposed into two cutting force components: F_x and F_y, which are shown in Fig. 2.12.

F_y is in the direction of the feed speed, and F_x is in the direction perpendicular to the feed speed. Therefore, the cutting components F_{xy}^{I} on deformed zone I can be decomposed using the equations given below:

$$F_x^{\mathrm{I}} = F_{xy}^{\mathrm{I}}\cos\kappa_r \tag{2.36}$$

$$F_y^{\mathrm{I}} = F_{xy}^{\mathrm{I}}\sin\kappa_r \tag{2.37}$$

Because the tools have a nose radius, their influence was considered using the concept of equivalent cutting edge. With this method, the straight and round parts of the cutting edge are replaced by a single straight cutting edge, i.e., the equivalent cutting edge regarded as the line joining the extreme points of the engaged cutting edge, as suggested by Colwell [77]. When the cut depth a_p is greater than the tool nose radius r_ε, the effective approach angle $\kappa_{r\varepsilon}$ is shown in Fig. 2.13 and given by Eq. 2.38.

$$\kappa_{r\varepsilon} = \cot^{-1}\left(\frac{r_\varepsilon tg(\kappa_r/2) + f/2}{a_p} + \cot\kappa_r\right) \tag{2.38}$$

Similarly, when the cut depth a_p is smaller than the tool nose radius r_ε, the effective approach angle $\kappa_{r\varepsilon}$ is given by

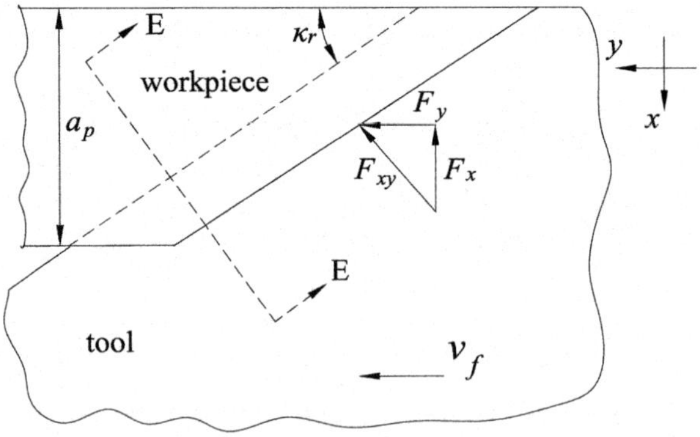

Fig. 2.12 Cutting components F_x and F_y

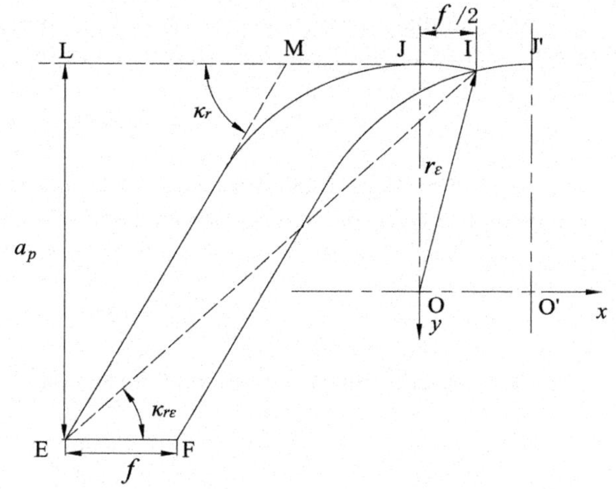

Fig. 2.13 Effective approach angle $\kappa_{r\varepsilon}$

$$\kappa_{r\varepsilon} = \cot^{-1}\left(\frac{\left(2r_\varepsilon a_p - a_p^2\right)^{1/2} + f/2}{a_p}\right) \quad (2.39)$$

(2) Analysis of cutting force components from deformed zone II.

The SiC particles will be extruded from the aluminum matrix, or fractured under the load of force $F_{\tau s}$ from the tool. When the particles roll on the rake face of the tool with the chip movement, the normal force F_n will be exerted on these rolled particles, thus inducing the rolled friction force. This scenario is described in Fig. 2.14.

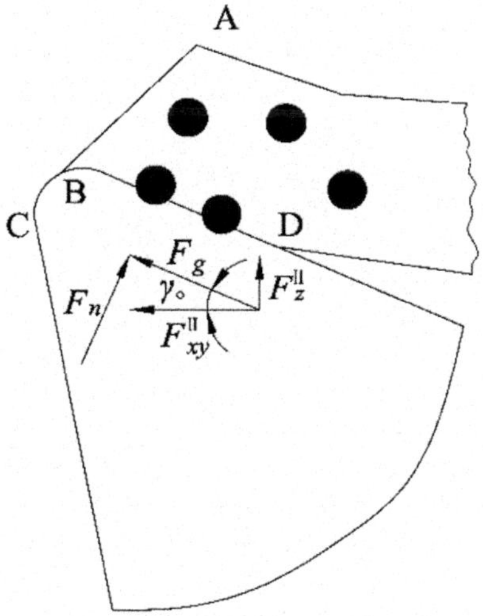

Fig. 2.14 Cutting force F_g in the deformed zone II

F_n is the normal force obtained by Eq. 2.27. F_g is the rolled friction force between the SiC particle and the rake face, which is given by [78]

$$F_g = K_g F_n N_A \tag{2.40}$$

where F_g is the rolling friction force on tool-chip interface, K_g is the roll friction coefficient, and N_A is the number of rolled SiC particles. N_A can be obtained by

$$N_A = \frac{\rho A}{\pi d_p^2} \tag{2.41}$$

The cut area A includes two different areas A_A and A_B. Therefore,

$$A = A_{\mathrm{A}} + A_{\mathrm{B}} \tag{2.42}$$

When the cut depth a_p is greater than the tool nose radius r_ε, zones A and B are as shown in Fig. 2.15. Zone A consists of straight lines EH, HG, GF, and FE. Zone B consists of circular arc HJI, circular arc IG, and straight line GH. The cut areas A_{A} and A_{B} can be given by Eq. 2.43 and Eq. 2.44 [79, 80], respectively:

$$A_{\mathrm{A}} = f\left[a_p - r_\varepsilon(1 - \cos\kappa_r)\right] - \frac{1}{4} f^2 \sin(2\kappa_r) \tag{2.43}$$

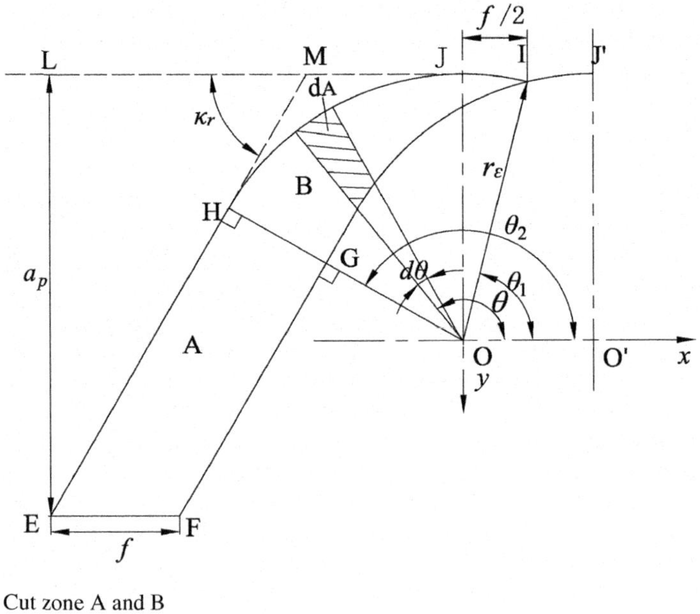

Fig. 2.15 Cut zone A and B

$$A_{\mathrm{B}} = \int_{\theta_1}^{\theta_2} r_\varepsilon - f\cos\theta - \left(r_\varepsilon^2 - f^2\sin^2\theta\right)^{1/2} d\theta \tag{2.44}$$

where θ, θ_1, and θ_2 are shown in Fig. 2.15. The angles θ_1, θ_2 can be expressed respectively by

$$\theta_1 = \cos^{-1}\left(\frac{f}{2r_\varepsilon}\right) \tag{2.45}$$

$$\theta_2 = \pi - \left(\frac{\pi}{2} - k_r\right) = \frac{\pi}{2} + \kappa_r \tag{2.46}$$

Similarly, when the cut depth a_p is smaller than the tool nose radius r_ε, A_{B} can be obtained used Eq. 2.44. A_{A} can be expressed by

$$A_{\mathrm{A}} = \int_{\theta_2}^{\theta_3} r_\varepsilon - (r_\varepsilon - a_p) * \csc\theta d\theta \tag{2.47}$$

Angle θ_2 and θ_3 can be expressed by

$$\theta_2 = \pi - \tan^{-1}\left(\frac{r - a_p}{\sqrt{2ra_p - d^2} - f}\right) \tag{2.48}$$

$$\theta_3 = \pi - \sin^{-1}\left(\frac{r_\varepsilon - a_p}{r_\varepsilon}\right) \tag{2.49}$$

The roll friction coefficient K_g affecting the roll friction force on the tool-chip interface is associated with the basic nature of tools and SiCp/Al composites, which can be obtained by [81]

$$K_g = \left(\frac{L}{4\pi} + \frac{\sigma_b + \sigma_s/2}{6H_B}\right) * L \tag{2.50}$$

In the equation above, L is the width of the groove formed on the rake face by SiC particles, which is shown in Fig. 2.16.

The width of the groove formed on the tool face by the SiC particles is associated with the nature of the tool material and particle material. The value of L can be obtained by [81]

$$L = \frac{1}{2}\left[\frac{3}{8}d_p\left(\frac{1 - v_1^2}{E_1} + \frac{1 - v_2^2}{E_2}\right)F_n\right]^{\frac{1}{3}} \tag{2.51}$$

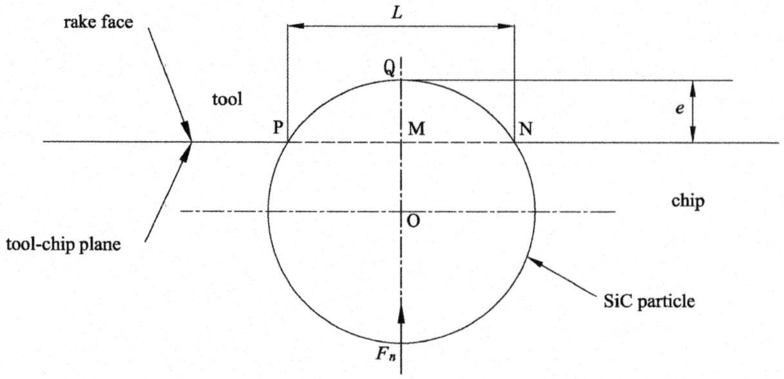

Fig. 2.16 Groove formed on rake face by SiC

The rolled friction force on the rake face by SiC particles can be decomposed into two components, which is given by

$$F_z^{\mathrm{II}} = F_g \sin \gamma_0 \tag{2.52}$$

$$F_{xy}^{\mathrm{II}} = F_g \cos \gamma_0 \tag{2.53}$$

Similarly, the cutting components F_{xy}^{II} on deformed zone II can be decomposed using the equations given below:

$$F_x^{\mathrm{II}} = F_{xy}^{\mathrm{II}} \cos \kappa_{r\varepsilon} \tag{2.54}$$

$$F_y^{\mathrm{II}} = F_{xy}^{\mathrm{II}} \sin \kappa_{r\varepsilon} \tag{2.55}$$

(3) Analysis of cutting force components from deformed zone III.

Owing to the cutting edge radius, a plowing force is important for the shear force. Because the particle fracture is considered separately, only the matrix aluminum is assumed to be subjected to plowing. To determine the force components owing to aluminum matrix plowing, the equations given in Ref. [82] that are based on a slip line field model for a rigid wedge sliding on a half-space and for orthogonal cutting can be used.

With the model above, the plowing force components on the cambered surface BC (in Fig. 2.9) is as shown in Fig. 2.17.

The plowing force acting on the tool in the radial direction of the cutting edge arc and through the middle of the arc, F_p, and that in the tangential direction of the cutting edge arc and through the middle of the arc, F_t, can be determined as follows [83]:

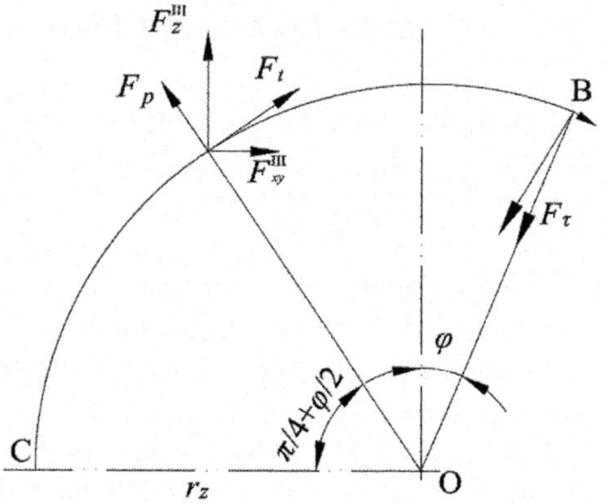

Fig. 2.17 Cutting force on the deformed zone III

$$F_p = \tau_s \cos(\varphi - \gamma_0) r_z \left(\frac{\pi}{2} + \varphi\right) d \tag{2.56}$$

$$F_t = \tau_s \sin(\varphi - \gamma_0) r_z \left(\frac{\pi}{2} + \varphi\right) d \tag{2.57}$$

Similarly, the plowing force on the cutting edge cambered surface can be decomposed into two components as follows:

$$F_z^{\mathrm{III}} = F_p \sin\left(\frac{\pi}{4} + \frac{\varphi}{2}\right) - F_t \cos\left(\frac{\pi}{4} + \frac{\varphi}{2}\right) \tag{2.58}$$

$$F_{xy}^{\mathrm{III}} = F_p \cos\left(\frac{\pi}{4} + \frac{\varphi}{2}\right) - F_t \sin\left(\frac{\pi}{4} + \frac{\varphi}{2}\right) \tag{2.59}$$

Similarly, the cutting force components F_{xy}^{III} on deformed zone II can be decomposed using the equations given below:

$$F_x^{\mathrm{III}} = F_{xy}^{\mathrm{III}} \cos \kappa_{r\varepsilon} \tag{2.60}$$

$$F_y^{\mathrm{III}} = F_{xy}^{\mathrm{III}} \sin \kappa_{r\varepsilon} \tag{2.61}$$

(4) Total cutting component.

The total cutting force components along the X, Y, and Z directions are given by

$$F_z = F_z^{\mathrm{I}} + F_z^{\mathrm{II}} + F_z^{\mathrm{III}} \tag{2.62}$$

$$F_x = F_x^{\mathrm{I}} + F_x^{\mathrm{II}} + F_x^{\mathrm{III}} \tag{2.63}$$

$$F_y = F_y^{\mathrm{I}} + F_y^{\mathrm{II}} + F_y^{\mathrm{III}} \tag{2.64}$$

2.4.2 Experiment and Discussion

The ranges of cutting conditions selected were based on literature [47–49]. Accordingly, the selected values are as shown in Table 2.15.

Cutting experiments under the same cutting conditions were performed five times to minimize the effects of the abnormal disturbance. The software can monitor and store force signals online during the machining process. The cutting force values can be saved as Microsoft Excel files in accordance with the three force components for subsequent analysis. The average force for each cutting condition was calculated and shown in Table 2.16.

During each experiment, chip samples were collected for measuring the chip thickness using a micrometer, as shown in Fig. 2.18. The average cutting temperature was measured using an infrared thermal imager FLUKE TI450, as shown in Fig. 2.19. These thickness and temperature values are shown in Table 2.17.

The comparison between the predicted and experimental force results is presented in Table 2.18.

It is observed that the predicted value of the resultant cutting force $F_{\tau s}$ agrees well with the experimental value. The predicted deviation is less than 4%. The maximum predicted deviation is − 3.53% on the condition that the tool nose radius is 0.8 mm, feed rate is 0.05 mm/rev, cutting speed is 120 mm/s, and cutting depth is 0.5 mm. The minimum predicted deviation is − 0.02% on the condition that the tool nose radius

Table 2.15 Cut parameters

No	Tool nose radius r_ε (mm)	Feed rate f (mm/rev)	Cutting speed v_c (m/min)	Cutting depth a_p (mm)
1	0.6	0.05	120	1
2	0.6	0.2	120	1
3	0.6	0.05	120	0.5
4	0.6	0.2	120	0.5
5	0.6	0.05	40	1
6	0.6	0.2	40	1
7	0.6	0.05	40	0.5
8	0.6	0.2	40	0.5
9	0.8	0.05	120	1
10	0.8	0.2	120	1
11	0.8	0.05	120	0.5
12	0.8	0.2	120	0.5
13	0.8	0.2	40	1
14	0.8	0.05	40	1
15	0.8	0.05	40	0.5
16	0.8	0.2	40	0.5

Table 2.16 Experimental cutting force components

No	Force component F_z (N)	Force component F_x (N)	Force component F_y (N)	Cutting force $F_{\tau s}$ (N)
1	58.65	18.91	30.74	68.87
2	183.15	50.96	63.46	200.42
3	31.31	16.03	14.00	37.86
4	90.07	29.22	22.10	97.24
5	58.06	21.26	30.38	68.89
6	175.84	45.40	55.99	190.04
7	32.07	16.41	14.32	38.77
8	97.77	31.53	23.23	105.32
9	64.59	25.05	32.87	76.68
10	192.92	60.88	70.42	214.21
11	34.13	20.60	14.75	42.51
12	108.09	41.41	29.04	119.33
13	196.43	58.58	68.24	216.04
14	65.99	26.15	35.68	79.45
15	36.04	22.34	15.97	45.31
16	108.43	42.13	28.45	119.75

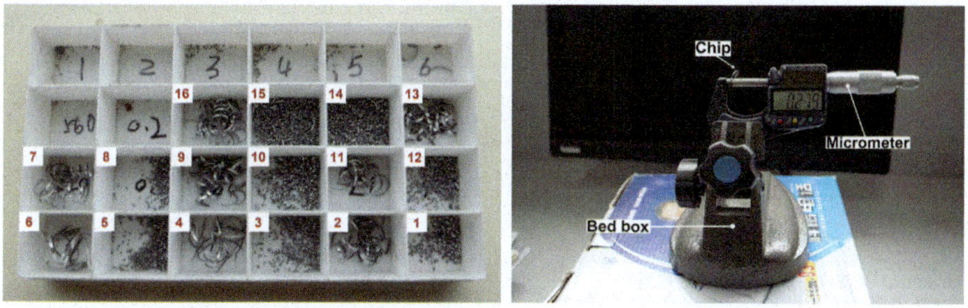

Fig. 2.18 Measurement of chip thickness

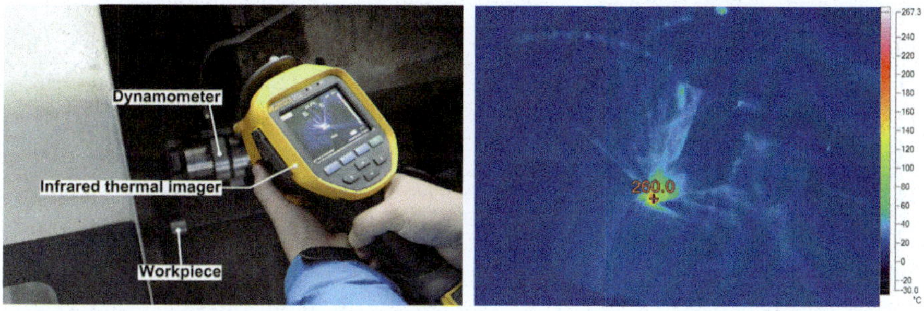

Fig. 2.19 Measurement of cutting temperature

Table 2.17 Chip thickness and cutting temperature

No	Cutting temperature T (°C)	Chip thickness b_c (mm)
1	225.70	0.074
2	254.00	0.259
3	208.00	0.053
4	238.30	0.180
5	211.60	0.049
6	249.10	0.237
7	199.30	0.052
8	227.60	0.181
9	238.2	0.080
10	265.60	0.279
11	216.10	0.052
12	241.2	0.197
13	250.2	0.256
14	218.8	0.078
15	194.9	0.052
16	238.5	0.194

Table 2.18 Predicted cutting force and deviation

No	Predicted cutting force (N)				Cutting force deviation			
	F_z	F_x	F_y	$F_{\tau s}$	F_z (%)	F_x (%)	F_y (%)	$F_{\tau s}$ (%)
1	59.99	20.28	26.92	68.81	2.28	7.24	− 12.44	− 0.08
2	184.27	49.78	60.09	200.11	0.61	− 2.33	− 5.31	− 0.15
3	31.53	14.70	11.92	36.77	0.70	− 8.29	− 14.86	− 2.87
4	93.68	27.92	20.18	99.82	4.01	− 4.47	− 8.66	2.65
5	60.37	19.84	26.34	68.79	3.98	− 6.69	− 13.30%	− 0.15
6	176.80	43.60	52.63	189.56	0.55	− 3.97	− 5.99	− 0.25
7	32.47	15.09	12.24	37.84	1.22	− 8.01	− 14.56	− 2.41
8	98.95	29.85	21.58	105.58	1.21	− 5.33	− 7.12	0.25
9	64.77	26.55	29.28	75.88	0.28	5.99	− 10.93	− 1.05
10	193.99	64.05	65.23	214.46	0.55	5.20	− 7.37	0.12
11	34.23	18.91	12.33	41.00	0.30	− 8.22	− 16.38	− 3.53
12	110.45	44.74	26.58	122.10	2.19	8.05	− 8.46	2.31
13	197.37	61.48	62.62	216.00	0.48	4.94	− 8.24	− 0.02
14	69.72	28.39	31.31	81.53	5.65	8.56	− 12.25	2.62
15	37.75	20.90	13.63	45.25	4.73	− 6.43	− 14.62	− 0.13
16	110.76	44.47	26.42	122.25	2.15	5.56	− 7.15	2.08
Average value of deviation					1.93	6.20	10.48	1.29

is 0.8 mm, feed rate is 0.2 mm/rev, cutting speed is 40 mm/s, and cutting depth is 1 mm. The average predicted deviation is 1.29%, indicating that the predicted resultant cutting force agreed well with the experimental cutting force.

Next, the predicted value of the cutting force component F_z also agrees well with the experimental value. The predicted deviation is less than 6%. The maximum predicted deviation is – 5.65% on the condition that the tool nose radius is 0.8 mm, feed rate is 0.05 mm/rev, cutting speed is 40 mm/s, and cutting depth is 1 mm. The minimum predicted deviation is 0.28% on the condition that the tool nose radius is 0.8 mm, feed rate is 0.05 mm/rev, cutting speed is 120 mm/s, and cutting depth is 1 mm. The average predicted deviation is 1.93%. Unlike the resultant cutting force $F_{\tau s}$, all the predicted values of the cutting force component F_z are higher than the experimental value.

Subsequently, for the cutting force components, i.e., both F_x and F_y, the predicted deviation is higher than the cutting force component F_z, In particular, the maximum predicted deviation of the cutting force component F_y is 16.38% on the condition that the tool nose radius is 0.8 mm, feed rate is 0.05 mm/rev, cutting speed is 120 mm/s, and cutting depth is 0.5 mm, at which the maximum predicted deviation of the resultant cutting force also occurs.

2.4.3 Conclusions

A mathematical model was developed for predicting the cutting forces in machining SiCp/Al composite. The cutting force was considered to be from three deformed zones: shear deformed zone, friction deformed zone, and plow deformed zone. The following conclusions can be drawn from this work:

1. A mathematical model based on the geometrical and mechanical nature of the tool and workpiece considering the effect of SiC reinforcement particles has been developed to predict the cutting force components in machining SiCp/Al composite.
2. The cutting force consisting of shear force, plowing force, and friction force was decomposed into three cutting force components to facilitate the analysis and calculation in the three vertical directions: *x*, *y*, and *z*.
3. Several cut experiments with various combinations of tool nose radii, feed rates, cutting speeds, and cutting depths have been conducted to verify the validity of the developed mathematical model of the cutting force. The cutting force, cutting temperature, and chip thickness have been measured.
4. The predicted cutting force components agreed well with the measured cutting force components in most cutting conditions.
5. The proposed geometrical and mechanical modeling approach could achieve the early, inexpensive, and exact prediction of the cutting force in machining SiCp/Al composite.

References

1. Rawal S (2014) Metal matrix composites for space applications. J Mater 53:14–17
2. Rohatgi P (1991) Cast aluminum-matrix composite for automotive applications. J Mater 43:10–15
3. Narutaki N (1996) Machining of MMCs. VDI BERICHTE NR. 1276:359–370
4. Chinmaya RD, Yung CS (2012) Modeling of machining of composite materials: a review. Int J Mach Tools Manuf 57:102–121
5. Sekhar R, Singh TP (2015) Mechanisms in turning of metal matrix composites: a review. J Mater Res Technol 4:197–207
6. Kumar R, Chauhan S (2015) Study on surface roughness measurement for turning of Al 7075/10/SiCp and Al 7075 hybrid composites by using response surface methodology (RSM) and artificial neural networking (ANN). Measurement 65:166–180
7. Zhou L, Wang Y, Ma ZY, Yu XL (2014) Finite element and experimental studies of the formation mechanism of edge defects during machining of SiCp/Al composites. Int J Mach Tools Manuf 84:9–16
8. Yingfei G, Jiuhua X, Hui Y (2010) Diamond tools wear and their applicability when ultra-precision turning of SiCp/2009Al matrix composite. Wear 269:699–708
9. Kannan S, Kishawy HA, Deiab I (2009) Cutting forces and TEM analysis of the generated surface during machining metal matrix composites. J Mater Process Technol 209:2260–2269
10. Priyadarshi D, Sharma RK (2016) Effect of type and percentage of reinforcement for optimization of the cutting force in turning of aluminium matrix nanocomposites using response surface methodologies. J Mech Sci Technol 30:1095–1101
11. Aurich JC, Zimmermann M, Schindler S, Steinmann P (2016) Turning of aluminum metal matrix composites: influence of the reinforcement and the cutting condition on the surface layer of the workpiece. Adv Manuf 4:225–236
12. Ozben T, Kilickap E, Çakır O (2008) Investigation of mechanical and machinability properties of SiC particle reinforced Al-MMC. J Mater Process Technol 198:220–225
13. Canakci A, Varol T (2014) Microstructure and properties of AA7075/Al-SiC composites fabricated using powder metallurgy and hot pressing. Powder Technol 268:72–79
14. Canakci A, Arslan F, Varol T (2014) Physical and mechanical properties of stir-casting processed AA2024/B4Cp composites. Sci Eng Compos Mater 21:505–515
15. Huang ST, Zhou L, Yu XL, Cui Y (2011) Experimental study of high-speed milling of SiCp/Al composites with PCD tools. Int J Adv Manuf Technol 62:487–493
16. Singh J, Chauhan A (2014) Overview of aluminium matrix composites for automotive applications. Int J Appl Eng Res 9:959–966
17. Charles S, Arunachalam VP (2004) Property analysis and mathematical modeling of machining properties of aluminium alloy hybrid (Al-alloy/SiC/flyash) composites produced by liquid metallurgy and powder metallurgy techniques. Indian J Eng Mater S 11:473–480
18. Wang T, Xie LJ, Wang XB (2015) Simulation study on defect formation mechanism of the machined surface in milling of high volume fraction SiCp/Al composite. Int J Adv Manuf Technol 79:1185–1194
19. Seeman M, Ganesan G, Karthikeyan R, Velayudham A (2010) Study on tool wear and surface roughness in machining of particulate aluminum metal matrix composite-response surface methodology approach. Int J Adv Manuf Technol 48:613–624
20. Bhushan RK, Kumar S, Das S (2010) Effect of machining parameters on surface roughness and tool wear for 7075 Al alloy SiC composite. Int J Adv Manuf Technol 50:459–469
21. Ekici E, Samtaş G, Gülesin M (2014) Experimental and Statistical Investigation of the Machinability of Al-10% SiC MMC produced by hot pressing method. Arab J Sci Eng 39:3289–3298
22. Tamang SK, Chandrasekaran M (2015) Modeling and optimization of parameters for minimizing surface roughness and tool wear in turning Al/SiCp MMC, using conventional and soft computing techniques. Adv Pro Eng Manag 10:59–72

23. Cheung CF, Lee WB (2000) A theoretical and experimental investigation of surface roughness formation in ultra-precision diamond turning. Int J Mach Tools Manuf 40:979–1002
24. Dogra M, Sharma VS, Dureja J (2011) Effect of tool geometry variation on finish turning—a review. J Eng Sci Technol Rev 4:1–13
25. Zhang SJ, To S, Wang SJ, Zhu ZW (2015) A review of surface roughness generation in ultra-precision machining. Int J Mach Tools Manuf 91:76–95
26. Baburaj E, Mohana Sundaram KM, Senthil P (2016) Effect of high speed turning operation on surface roughness of hybrid metal matrix (Al-SiCp-fly ash) composite. J Mech Sci Technol 30:89–95
27. Sawangsri W, Cheng K (2014) An innovative approach to cutting force modelling in diamond turning and its correlation analysis with tool wear. Proc IMechE Part B: J Eng Manuf 230(3):405–415
28. Kumar A, Mahapatra MM, Jha PK (2014) Effect of machining parameters on cutting force and surface roughness of in situ Al-4.5%Cu/TiC metal matrix composites. Measurement 48:325–332
29. Polini W, Turchetta S (2014) Cutting force, tool life and surface integrity in milling of titanium alloy Ti-6Al-4V with coated carbide tools. Proc IMechE Part B: J Eng Manuf 230(4):694–700
30. Kishawy HA, Hosseini A, Moetakef-Imani B, Astakhov VP (2012) An energy based analysis of broaching operation: cutting forces and resultant surface integrity. CIRP Ann Manuf Technol 61(1):107–110
31. Dandekar CR, Shin YC (2012) Modelling of machining of composite materials: a review. Int J Mach Tools Manuf 57:102–121
32. Jeyakumar S, Marimuthu K, Ramachandran T (2013) Prediction of cutting force, tool wear and surface roughness of Al6061/SiC composite for end milling operations using RSM. J Mech Sci Technol 27(9):2813–2822
33. Venkatesan K, Ramanujam R, Joel J, Jeyapandiarajan P, Vignesh M, Tolia DJ, Krishna RV (2014) Study of cutting force and surface roughness in machining of Al alloy hybrid composite and optimized using response surface methodology. Procedia Eng 97:677–686
34. Shoba CH, Ramanaiah N, Rao DN (2015) Effect of reinforcement on the cutting forces while machining metal matrix composites—an experimental approach. Eng Sci Technol Int J 18(4):658–663
35. Subramanian AVM, Nachimuthu MDG, Cinnasamy V (2017) Assessment of cutting force and surface roughness in LM6/SiCp using response surface methodology. J Appl Res Technol 15(3):283–296
36. Aezhisai Vallavi MS, Mohan Das Gandhi N, Velmurugan C (2018) Application of genetic algorithm in optimisation of cutting force of Al/SiCp metal matrix composite in end milling process. Int J Mater Prod Technol 56(3):234–252
37. Pramanik A, Zhang LC, Arsecularatne JA (2007) An FEM investigation into the behavior of metal matrix composites: tool-particle interaction during orthogonal cutting. Int J Mach Tools Manuf 47(10):1497–1506
38. Zhou L, Huang ST, Wang D, Yu XL (2011) Finite element and experimental studies of the cutting process of SiCp/Al composites with PCD tools. Int J Adv Manuf Technol 52(5–8):619–626
39. Fathipour M, Hamedi M, Yousefi R (2013) Numerical and experimental analysis of machining of Al (20 vol% SiC) composite by the use of ABAQUS software. Materialwiss Werkst 44(1):14–20
40. Doomra VK, Debnath K, Singh I (2014) Drilling of metal matrix composites: Experimental and finite element analysis. Proc IMechE Part B J Eng Manuf 229(5):886–890
41. Chen X, Xie L, Xue X, Wang X (2017) Research on 3D milling simulation of SiCp/Al composite based on a phenomenological model. Int J Adv Manuf Technol 92(5–8):2715–2723
42. Kishawy HA, Kannan S, Balazinski M (2004) An energy based analytical force model for orthogonal cutting of metal matrix composites. CIRP Ann Manuf Technol 53(1):91–94
43. Du J, Li J, Yao Y, Hao Z (2014) Prediction of cutting forces in mill-grinding SiCp/Al composites. Mater Manuf Proc 29(3):314–320

44. Campocasso S, Costes J-P, Fromentin G, Bissey-Breton S, Poulachon G (2015) A generalised geometrical model of turning operations for cutting force modelling using edge discretisation. Appl Math Model 39(21):6612–6630
45. Weng J, Zhuang K, Chen D, Guo S, Ding H (2017) An analytical force prediction model for turning operation by round insert considering edge effect. Int J Mech Sci 128–129:168–180
46. Okafor C, Sultan AA (2016) Development of a mechanistic cutting force model for wavy-edge bull-nose helical end-milling of Inconel 718 under emulsion cooling strategy. Appl Math Model 40(4):2637–2660
47. Pramanik A, Zhang LC, Arsecularatne JA (2006) Prediction of cutting forces in machining of metal matrix composites. Int J Mach Tools Manuf 46(14):1795–1803
48. Dabade UA, Dapkekar D, Joshi SS (2009) Modelling of chip-tool interface friction to predict cutting forces in machining of Al/SiCp composites. Int J Mach Tools Manuf 49(9):690–700
49. Sikder S, Kishawy HA (2012) Analytical model for force prediction when machining metal matrix composite. Int J Mech Sci 59(1):95–103
50. Lu X, Zhang H, Jia Z, Feng Y, Liang SY (2018) Cutting parameters optimization for MRR under the constraints of surface roughness and cutter breakage in micro-milling process. J Mech Sci Technol 32:3379–3388
51. Leo Kumar SP (2019) Measurement and uncertainty analysis of surface roughness and material removal rate in micro turning operation and process parameters optimization. Measurement 140:538–547
52. Zahoor S, Ameen F, Abdul-Kader W, Stagner J (2019) Environmentally conscious machining of Inconel 718: surface roughness, tool wear, and material removal rate assessment. Int J Adv Manuf Technol 106:303–313
53. Meenu SK (2019) Optimization of the surface roughness and material removal rate in turning of unidirectional glass fiber reinforced plastics using the fuzzy-grey relational technique. Indian J Eng Mater S 26:7–19
54. Soorya Prakash K, Gopal PM, Karthik S (2020) Multi-objective optimization using Taguchi based grey relational analysis in turning of rock dust reinforced aluminum MMC. Measurement 157
55. Beffort O, Long S, Cayron C, Kuebler J, Buffat PA (2007) Alloying effects on microstructure and mechanical properties of high volume fraction SiC-particle reinforced Al-MMCs made by squeeze casting infiltration. Compos Sci Technol 67:737–745
56. Ren S, He X, Qu X (2008) Effect of controlled interfacial reaction on the microstructure and properties of the SiCp/Al composites prepared by pressureless infiltration. J Alloys Comps 455:424–431
57. Huang S, Guo L, He H, Yang H, Su Y, Xu L (2018) Experimental study on SiCp/Al composites with different volume fractions in high-speed milling with PCD tools. Int J Adv Manuf Technol 97:2731–2739
58. Wang T, Xie LJ, Wang XB, Jiao L, Shen JW, Xu H, Nie FM (2013) Surface integrity of high speed milling of Al/SiC/65p aluminum matrix composites. Procedia CIRP 8:475–480
59. Wang T, Xie L, Wang X, Ding Z (2015) PCD tool performance in high-speed milling of high volume fraction SiCp/Al composites. Int J Adv Manuf Technol 78:1445–1453
60. Yang Q, Wu Z, Zhan L, Li N, He R (2015) Shrestha, An experimental study on milling of high-volume fraction SiCP/Al composites with PCD tools of different grain size. Int J Adv Manuf Technol 79:1699–1705
61. Wang J, Zuo J, Shang Z, Fan X (2019) Modeling of cutting force prediction in machining high-volume SiCp/Al composites. Appl Math Model 70:1–17
62. Duan C, Sun W, Che M, Yin W (2019) Effects of cooling and lubrication conditions on tool wear in turning of Al/SiCp composite. Int J Adv Manuf Technol 103:1467–1479
63. Peace GS (1992) Taguchi methods, a hands-on approach. Addison-Wesley MA
64. Mathews PG (2005) Design of experiments with minitab. ASQ Quality Press (2005)
65. Mia M, Dey PR, Hossain MS, Arafat MT, Asaduzzaman M, Shoriat Ullah M, Tareq Zobaer SM (2018) Taguchi S/N based optimization of machining parameters for surface roughness, tool wear and material removal rate in hard turning under MQL cutting condition. Measurement 122:380–391

66. Mia M, Dhar NR (2016) Optimization of surface roughness and cutting temperature in high-pressure coolant-assisted hard turning using Taguchi method. Int J Adv Manuf Technol 88:739–753
67. Dabade UA, Sonawane HA, Joshi SS (2010) Cutting forces and surface roughness in machining Al/SiCp composites of varying composition. Mach Sci Technol 14:258–279
68. Fan W, Zheng L, Ji W, Zhao X, Wang L, Yang Y (2019) Eddy current-based vibration suppression for finish machining of assembly interfaces of large aircraft vertical tail. J Manuf Sci Eng 141:071012
69. Wang J, Wu X, Fan X (2015) A two-stage ant colony optimization approach based on a directed graph for process planning. Int J Adv Manuf Technol 80:839–850
70. Tzeng C-J, Lin Y-H, Yang Y-K, Jeng M-C (2009) Optimization of turning operations with multiple performance characteristics using the Taguchi method and Grey relational analysis. J Mater Process Technol 209:2753–2759
71. Mia M, Bashir MA, Khan MA, Dhar NR (2016) Optimization of MQL flow rate for minimum cutting force and surface roughness in end milling of hardened steel (HRC 40). Int J Adv Manuf Technol 89:675–690
72. Mia M, Khan MA, Rahman SS, Dhar NR (2016) Mono-objective and multi-objective optimization of performance parameters in high pressure coolant assisted turning of Ti-6Al-4V. Int J Adv Manuf Technol 90:109–118
73. Moufki A, Devillez A, Dudzinski D, Molinari A (2004) Thermomechanical modelling of oblique cutting and experimental validation. Int J Mach Tools Manuf 44(9):971–989
74. Huang Y, Liang SY (2005) Modelling of cutting forces under hard turning conditions considering tool wear effect. J Manuf Sci Eng T ASME 127(2):262–270
75. Lin JT, Bhattacharyya D, Ferguson WG (1998) Chip formation in the machining of SiC-particle-reinforced aluminum-matrix composites. Compos Sci Technol 58(2):285–291
76. Lee EH, Shaffer BW (1951) The theory of plasticity applied to a problem of machining. J Appl Mech 15–20
77. Colwell LV (1954) Predicting the angle of chip flow for single-point cutting tool. T ASME 76:199–204
78. Jiang J, Sheng F, Ren F (1998) Modelling of two body abrasive wear under multiple contact conditions. Wear 217(1):35–45
79. Young HT, Mathew P, Oxley PLB (1987) Allowing for nose radius effects in predicting the chip flow direction and cutting forces in bar turning. Proc IMechE Part B: J Eng Manuf 201(3):213–226
80. Arsecularatne JA, Mathew P, Oxley PLB (1995) Prediction of chip flow direction and cutting forces in oblique machining with nose radius tools. Proc IMechE Part B: J Eng Manuf 209(4):303–315
81. Wang Z (1993) A mechanism of rolling friction and the coefficient of rolling friction. J Shanghai Inst Mech Eng 15(4):35–43
82. Waldorf DJ, DeVo RE, Kapoor SG (1998) A slip-line field for ploughing during orthogonal cutting. J Manuf Sci Eng -T ASME 120(4):693–699
83. Waldorf DJ (2006) A simplified model for ploughing forces in turning. J Manuf Proc 8(2):447–454

Chapter 3
Intelligent Optimization of Process Plans

3.1 Introduction

In the modern computer-integrated manufacturing system (CIMS), the CAPP system plays an important role [1]. Generally, process planning involves two activities: operation determining and operation sequencing. In CAPP systems, these two activities must be executed simultaneously to achieve a good solution. Thus far, some effort has been made to address this problem, for instance, by designing a more feasible mathematical model or developing a more efficient approach. The application of some artificial intelligence approaches in the PP problem has especially promoted the development of CAPP technology. These approaches can be categorized as the genetic algorithm (GA) [1, 2], simulated annealing (SA) [2, 3], tabu search (TS) [4, 5], ant colony optimization (ACO) [6–8], particle swarm optimization (PSO) [9–13], honey bees mating optimization [14], and hybrid approaches [2, 15].

In 1995, Kennedy and Eberhart proposed the PSO algorithm [16]. The PSO is a new swarm optimization approach that can optimize engineering problems in aspects such as turning process modelling [17], assembly sequence planning [18], and process parameter optimization (TSP) [19].

The search for applications of PSO in the PP problem was first introduced by Guo et al. [9]. The process plan particle encoded/decoded strategy and some modified operators were designed. Kafashi et al. [10] optimized the setup planning using cost indices based on constraints such as the TAD (tool approach direction), the tolerance relation between features, and the feature precedence relations. Wang et al. [11] proposed an innovative process plan representation in which operation selection and operation sequencing were introduced simultaneously in a particle. Two local search operators are incorporated into the traditional PSO to achieve the optimal solution. Li et al. [12] modified the traditional PSO for the process planning problem. Miljković et al. [13] adopted the AND/OR network representation to describe the flexibility of the machine, tool, TAD, process and sequence and a performed multi-objective

J. Wang, *Intelligent Manufacturing System and Intelligent Workshop*,
Advanced and Intelligent Manufacturing in China,
https://doi.org/10.1007/978-981-99-2011-2_3

optimization procedure for the minimization of the production time and production cost using a modified PSO algorithm on this representation.

Ant colony optimization algorithm, proposed by Dorigo et al. [20], is a new nature inspired meta-heuristics used for solving the NP-complete combinatorial optimization problem. Various optimization problems have been solved using ACO algorithm such as: TSP [20], vehicle routing problem (VRP) [21, 22], hole-making process optimization [23, 24], assembly line balancing problem [25, 26], job shop scheduling problem [27, 28], etc.

The search on application of ACO in PP problem was introduced first by Krishna and Mallikarjuna [29]. Krishna and Mallikarjuna proposed a novel approach to apply the ant colony algorithm as a global search technique for PP problem by considering various feasibility constraints. A relative cost matrix is generated taking into consideration the change of machine tool, cutting tool and machining parameters. This search discussed the potential of ACO in PP problem simply. Some deep discussion is lacked, for example, application of ACO considering the complex machining environments, the analysis of ACO in large-scale PP problem, etc. Liu et al. [30] converted the PP problem to be a constraint-based traveling salesman problem by a weight graph, and constructed a mathematical model for PP problem by considering process constraints and optimization objective. The constraint matrix and state matrix are used to show the state of the operations and the searching range of the candidate operations. Simulation results for the complex part are satisfactory, but a detail explanation for the adjusted strategy of ACO parameters is expected to convince the simulation results probably.

Besides PSO and ACO, many optimization approaches have been developed and widely applied for solving PP problem in the past two decades. Zhang et al. [1] proposed a GA for a novel CAPP model. Li et al. [2] incorporated an SA into a GA to improve its searching efficiency. Ma et al. [3] modeled the PP problem as a combinational optimization problem with constraints. An entire solution space is constructed in reference to precedence constraints among operations. An SA algorithm is then proposed to address the PP problem. Li et al. [4] applied a TS-based approach to address the PP problem. In this approach, a mapping relationship is established between process constraints among features and precedence constraints among operations. Lian et al. [5] proposed a multi-dimensional tabu search (MDTS) approach to address the PP problem. Some local search strategies for different parts have been integrated into this TS approach. Liu et al. [6] constructed a mathematical model for the PP problem by considering the process constraints and optimization objectives. The ACO approach has been developed to optimize the PP problem based on this mathematical model. Wen et al. [14] proposed a new method based on the honey bees mating optimization (HBMO) approach to optimize the PP problem. The solution encoding, crossover operator, and local search strategies were developed according to the characteristics of the PP problem. Huang et al. [15] designed a hybrid algorithm combining a graph and the GA to optimize process plans. The precedence constraints are mapped to an operation precedence graph on which an improved GA was applied to solve the PP problem.

3.2 Problem Modelling

In the PP problem, two tasks have to be performed, namely, operation determining and operation sequencing. For operation determining, the method of mapping from features to operations is widely used in the PP problem. The attributes of each feature determines the corresponding machining methods, which can be expressed by the alternative operations. The combination of machines, cutting tools, and tool approach directions (TAD) comprise all of the different operation types (OPTs) for a feature of a part, which will be selected to determine the final process plans [1, 11]. For an operation, there are a set of OPTs under which the operation can be executed.

For a part with *m* feathers, the relationships between part, feather and operation are shown in Fig. 3.1.

Accordingly, an integrated process plan can be represented as follows.

$$PP = \{OP_1, OP_2, \ldots, OP_i, \ldots, OP_n\} \tag{3.1}$$

$$OP_i = \left\{OPT_{i1}, OPT_{i2}, \ldots, OPT_{ij}, \ldots, OPT_{im}\right\} \tag{3.2}$$

$$OPT_{ij} = \left\{M_{ij}, T_{ij}, TAD_{ij}\right\} \tag{3.3}$$

OP_i is the *i*th operation, and OPT_{ij} is the *j*th alternative selection of the operation OP_i, M_{ij}, T_{ij} and TAD_{ij} are the ID of the selected machine, tool and *TAD* for operation OPT_{ij}, respectively.

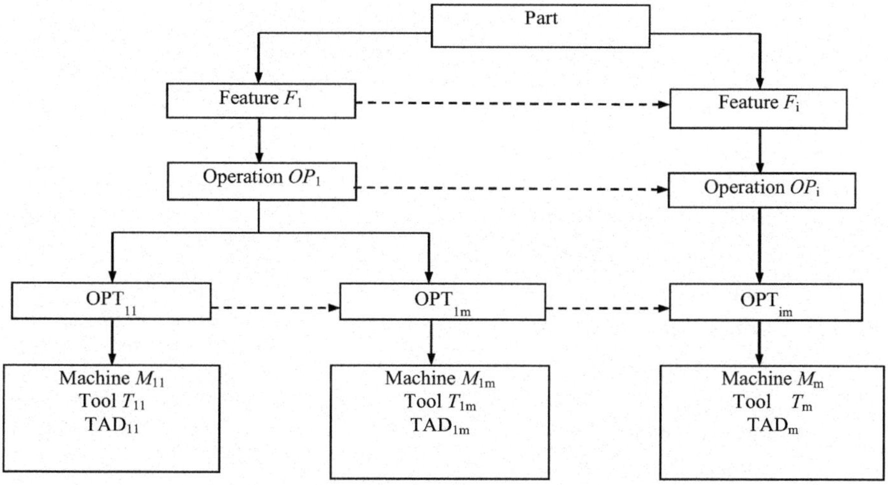

Fig. 3.1 The representation of process plan for a part with *m* feathers

An example part in Fig. 3.2 is used to demonstrate the representation of a process plan [7]. The machining details of the Part 1 are listed in Table 3.1. The candidate operations are listed in Table 3.2.

In addition to operation determining, operation sequencing is another task in the PP problem. The operations should be sequenced under the conditions of satisfying the precedence constraints among operations. In general, these precedence constraints are as following [4, 6, 12, 15].

(1) Primary surfaces prior to secondary surface.
(2) Planes prior to its associated features.
(3) Rough machining operation prior to finish machining operation.
(4) Datum surfaces prior to its associated features.
(5) Some good manufacturing practice.

According to the precedence constraints in Table 3.3, an available process plan for Part 1 is shown in Table 3.4.

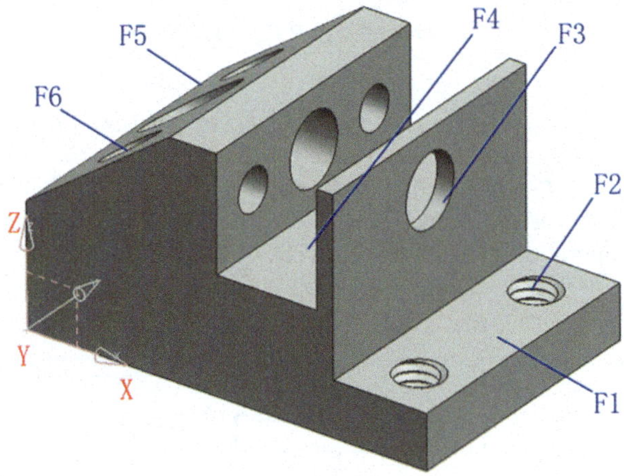

Fig. 3.2 An example part—Part 1

Table 3.1 Machining details for Part 1

Feathers	Operations	Machines	Tools	TADs	Description
F1	Milling(OP_1)	M2	T1	+ X, + Z	M1: Drilling press M2: Vertical milling machine T1: Milling cutter; T2:Drill1 T3: Tapping tool; T4: Drill2 T5: Reamer1; T6: Slot cutter T7: Chamfer cutter; T8: Drill3 T9: Reamer2
F2	Drilling(OP_2)	M1, M2	T2	− Z	
	Tapping(OP_3)		T3		
F3	Drilling(OP_4)	M1, M2	T4	− X	
	Reaming(OP_5)		T5		
F4	Milling(OP_6)	M2	T6	+ Z	
F5	Milling(OP_7)	M2	T7	+ Y, -Z	
F6	Drilling(OP_8)	M1, M2	T8	+ X	
	Reaming(OP_9)		T9		

Table 3.2 Candidate operations for Part 1

Operations	Candidate operations	Machines	Tools	TADs
OP_1	OPT_{11}	M1	T1	+ X
	OPT_{12}	M1	T1	+ Z
OP_2	OPT_{21}	M1	T2	− Z
	OPT_{32}	M2	T2	− Z
OP_3	OPT_{31}	M1	T3	− Z
	OPT_{32}	M2	T3	− Z
OP_4	OPT_{41}	M1	T4	− X
	OPT_{42}	M2	T4	− X
OP_5	OPT_{51}	M1	T5	− X
	OPT_{52}	M2	T5	− X
OP_6	OPT_{61}	M1	T6	+ Z
OP_7	OPT_{71}	M1	T7	− Z
	OPT_{72}	M1	T7	+ Y
OP_8	OPT_{81}	M1	T8	+ X
	OPT_{82}	M2	T8	+ X
OP_9	OPT_{91}	M1	T9	+ X
	OPT_{92}	M2	T9	+ X

Table 3.3 Constraints for Part 1

Operations	Precedence constraint description	Constraint types
OP_1	OP_1 is prior to OP_2 and OP_3	Hard
	OP_1 is prior to OP_4 and OP_5	Soft
OP_2	OP_2 is prior to OP_3	Hard
OP_4	OP_4 is prior to OP_5	Hard
OP_4, OP_5	OP_4 and OP_5 are prior to OP_6	Hard
OP_6	OP_6 is prior to OP_2 and OP_3	Hard
OP_8	OP_8 is prior to OP_9	Hard
OP_8, OP_9	OP_8 and OP_9 are prior to OP_7	Hard

Table 3.4 Available process plan for Part 1

Operation	Candidate operations	Machine	Tool	TAD
OP_1	OPT_{11}	M1	T1	+ X
OP_2	OPT_{22}	M2	T2	− Z
OP_3	OPT_{31}	M1	T3	− Z
OP_4	OPT_{41}	M1	T4	− X
OP_5	OPT_{52}	M2	T5	− X
OP_6	OPT_{61}	M1	T6	+ Z
OP_8	OPT_{81}	M1	T8	+ X
OP_9	OPT_{92}	M2	T9	+ X
OP_7	OPT_{71}	M1	T7	− Z

3.3 Process Plan Evaluation Criterion

Lots of process planning evaluation criterions have been proposed in the past literatures. The criterion of minimum production cost is generally used. The production cost evaluating process plans comprises six factors: machine processing cost (*MC*), tool processing cost (*TC*), machine change cost (*MCC*), tool change cost (*TCC*), set-up cost (*SCC*), and additional penalty cost (*APC*) [1, 2, 4, 17]. The calculation procedures of these cost factors are described in detail below.

(1) Total Machine cost (*TMC*)

$$TMC = \sum_{i=1}^{n} MC_i \tag{3.4}$$

where n is the total number of operations and MC_i is the machine cost of the ith machine for an operation, a constant for a specific machine.

(2) Total Tool Cost (*TTC*)

$$TTC = \sum_{i=1}^{n} TC_i \tag{3.5}$$

where TC_i is the tool cost of the ith tool for an operation, a constant for a specific machine.

(3) Total Machine change cost (*TMCC*)

$$TMCC = MCC * NMC \tag{3.6}$$

where *MCC* is considered to be the same for each machine change, *NMC* is number of machine change, which can be calculated by Eqs. (3.7) and (3.8).

$$NMC = \sum_{i=1}^{n-1} \Omega_1(M_{i+1}, M_i) \tag{3.7}$$

$$\Omega_1(x, y) = \begin{cases} 1 & x \neq y \\ 0 & x = y \end{cases} \tag{3.8}$$

where M_i is the machine for the ith operation.

(4) Total tool change cost (*TTCC*).

$$TTCC = TCC * NTC \tag{3.9}$$

where *TCC* is considered to be the same for each machine change, *NTC* is the number of tool change, which can be calculated by Eqs. (3.8), (3.10) and (3.11).

$$NTC = \sum_{i=1}^{n-1} \Omega_2(\Omega_1(M_{i+1}, M_i), \Omega_1(T_{i+1}, T_i)) \tag{3.10}$$

$$\Omega_2(x, y) = \begin{cases} 0 & x = y = 0 \\ 1 & \text{otherwise} \end{cases} \tag{3.11}$$

where T_i is the tool for the ith operation.

(5) Total set-up cost (*TSCC*)

$$TSCC = SCC * NS \tag{3.12}$$

where *SCC* is considered to be the same for each set-up, *NS* is the number of set-ups, which can be calculated by Eqs. (3.13) and (3.14).

$$NS = NSC + 1 \tag{3.13}$$

$$NSC = \sum_{i=1}^{n-1} \Omega_2(\Omega_1(M_{i+1}, M_i), \Omega_1(TAD_{i+1}, TAD_{ii})) \tag{3.14}$$

where TAD_i is the TAD for the ith operation.

(6) Total additional penalty cost (*TAPC*)

$$TAPC = APC * NPC \tag{3.15}$$

where *APC* is the fixed penalty cost, *NPC* is number of violating constraints, which can be calculated by Eqs. (3.16) and (3.17).

$$NPC = \sum_{i=1}^{n-1} \sum_{j=i+1}^{n} \Omega_3(OP_i, OP_j) \tag{3.16}$$

$$\Omega_3(x, y) = \begin{cases} 1 \text{ The sequence of } x \text{ and } y \text{ operations violates constraints} \\ 0 \text{ The sequence of } x \text{ and } y \text{ operations meets constraints} \end{cases} \tag{3.17}$$

The definition of machine change, tool change, setup change has been illustrated in detail [1, 3]. In this paper, the combination of *TWC*, *TTC*, *TMCC*, *TTCC*, *TSCC*, and *TAPC* will be used as the objective of process planning problem, which can be defined as Total Production Costs (*TPC*) and calculated by Eq. (3.18).

$$\begin{aligned} TPC = {} & w_1 * TMC + w_2 * TTC + w_3 * TMCC \\ & + w_4 * TSCC + w_5 * TTCC + w_6 * TAPC \end{aligned} \tag{3.18}$$

In Eq. (3.18), w_1, w_2, w_3, w_4, w_5, and w_6 are weights of *TMC*, *TTC*, *TMCC*, *TTCC*, *TSCC*, and *TAPC* respectively, the value of which is limited in $\{0, 1\}$. These weights can be assigned referring to the active situations, which provides the flexibility to customize the optimization objective function according to various situations [3]. w_6 can be used to switch the penalty function of the 'soft' constraints on or off. If no 'soft' constraints are considered, it is assigned as 0. Otherwise, it is 1.

Set the costs for the Part 1 in Fig. 3.2 as shown in Table 3.5. Accordingly, the costs of the process plan in Table 3.4 can be calculated in Table 3.6.

Table 3.5 The costs for the Part 1

MC		*TC*									*MCC*	*TCC*	*SCC*
M1	M2	T1	T2	T3	T4	T5	T6	T7	T8	T9			
40	10	10	3	7	3	8	10	10	3	8	300	60	20

Table 3.6 The costs of the process plan in Table 3.4

Process plan	Operations	Candidate operations	Machine	Tool	TAD	*MC*	*TC*	*MCC*	*TCC*	*SCC*
1 ↓ 2 ↓ 3 ↓ 4 ↓ 5 ↓ 6 ↓ 8 ↓ 9 ↓ 7	OP_1	OPT_{11}	M1	T1	+X	40	10			
	OP_2	OPT_{22}	M2	T2	−Z	10	3	300	60	20
	OP_3	OPT_{31}	M1	T3	−Z	40	7	300	60	20
	OP_4	OPT_{41}	M1	T4	−X	40	3		60	20
	OP_5	OPT_{52}	M2	T5	−X	10	8	300	60	20
	OP_6	OPT_{61}	M1	T6	+Z	40	10	300	60	20
	OP_8	OPT_{81}	M1	T8	+X	40	3		60	20
	OP_9	OPT_{92}	M2	T9	+X	10	8	300	60	20
	OP_7	OPT_{71}	M1	T7	−Z	40	10	300	60	20
Subtotal						270	62	1800	480	180
Subtotal						332		2460		
Total						2792				

3.4 Solving Process Planning Problem Using a ACO Algorithm

3.4.1 Graph-Based Representation of Process Plan

To construct a feasible process plan with the ACO approach, the process planning problem has to be visualized and represented by a disjunctive weighted directed graph. The disjunctive graph is denoted as $D = (O, A, B)$, where O is a set of nodes, A is a set of directed arcs, and B is a set of undirected arcs. The nodes of O stand for all of the operations $\mathrm{Op_i}$, A corresponds to the precedence relationships between the operations of the parts. B represents the set of arcs connecting all possible combination of the nodes. Both A and B represent possible paths for ants travelling from one node to another. The ants are basically free to travel along the paths unless there is a precedence constraint specified by A. Figure 3.3 is the disjunctive graph for the Part 1 in Fig. 3.2.

Nodes O_1–O_{17} in Fig. 3.3 represent the candidate process set of 9 processes, 8 undirected arcs are used to connect nodes O_1, O_2, O_{12}, O_{13}, O_{14}, O_{15}, O_{16}, O_{17}, and the other 8 directed arcs are used to connect O_3, O_4, O_5, O_6, O_7, O_8, O_9, O_{10}. The OR node represents the alternative process set *OPT* of a certain process *OP*. For example, when O_2 under OR1 is selected as OP_2, then O_1 will be ignored.

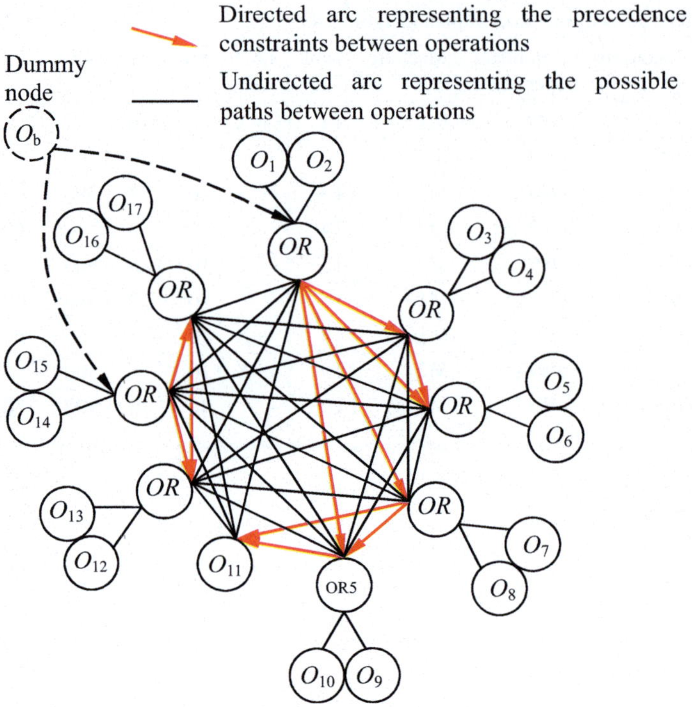

Fig. 3.3 A disjunctive weighted directed graph for the Part 1

The correspondence between process nodes and candidate process sets is shown in Table 3.7.

While applying the ACO in the process planning by the graph, the ant colony will be placed on the initial node visited by the ant colony first. The initial node determines which operation can be executed first. For the graph in Fig. 3.2, the nodes O_1, O_2, O_{14}, and O_{15} are likely to be selected as the initial source node, since operations OP_1 and OP_8 have no precedence operations. To facilitate the execution of ACO in process planning, a dummy node O_b acting as the initial node is added to connect the possibly executed operations first in the graph. The initial node O_b is used to connect the nodes O_1, O_2, O_{14}, and O_{15}.

3.4.2 Two-Stage ACO Algorithm

Selecting operations and sequencing operations have to be done in PP problem. Accordingly, the two-stage ACO approach is to deal with the operation selection and sequencing in two consecutive stages. The first stage corresponds to operation selection. Ant colony is deployed on the initial node of the graph, and ants traversed

Table 3.7 Nodes, Operations and candidate operations for Part 1

Nodes	Operations	Candidate operations	Machine	Tool	TAD
O_1	OP_1	OPT_{11}	M1	T1	+X
O_2		OPT_{12}	M1	T1	+Z
O_3	OP_2	OPT_{21}	M1	T2	−Z
O_4		OPT_{22}	M2	T2	−Z
O_5	OP_3	OPT_{31}	M1	T3	−Z
O_6		OPT_{32}	M2	T3	−Z
O_7	OP_4	OPT_{41}	M1	T4	−X
O_8		OPT_{42}	M2	T4	−X
O_9	OP_5	OPT_{51}	M1	T5	−X
O_{10}		OPT_{52}	M2	T5	−X
O_{11}	OP_6	OPT_{61}	M1	T6	+Z
O_{12}	OP_7	OPT_{71}	M1	T7	−Z
O_{13}		OPT_{72}	M1	T7	+Y
O_{14}	OP_8	OPT_{81}	M1	T8	+X
O_{15}		OPT_{82}	M2	T8	+X
O_{16}	OP_9	OPT_{91}	M1	T9	+X
O_{17}		OPT_{92}	M2	T9	+X

the necessary nodes to identify the preferred nodes set with the smaller *TPC*. The character of this stage is that the nodes of the graph are pheromone carriers. The second stage corresponds to operation sequencing, the unselected nodes in the first stage will be ignored completely in subsequent ant visits and the ants will deposit pheromones at the arcs with the aim of identifying the most favorable visited path. The arcs among nodes are pheromone carriers in this stage, which is different from the first stage. Procedures of the two stages are outlined in Fig. 3.4, and explanations for symbols used are listed in Table 3.8.

1. **Stage 1—Operation selection**

Ant colony K is deployed at the dummy start node O_b initially. The ants traverse all of the necessary nodes in accordance with the precedence constraints, until all the required operations are finished and a process plan is generated. Due to the alternative operations, not all the nodes in the weight graph have to be visited by the ants. After a node is visited by an ant, all the alternative nodes belonging to the same operation will be ignored. In this stage, nodes of the weight graph are the pheromone carriers. To choose the next node v among all the possible operation nodes connecting to the current node u, an ant k is guided by a pheromone amount τ_v and a heuristic information η_v. The heuristic information η_v is calculated as following.

$$\eta_v = \frac{E}{PC} \tag{3.19}$$

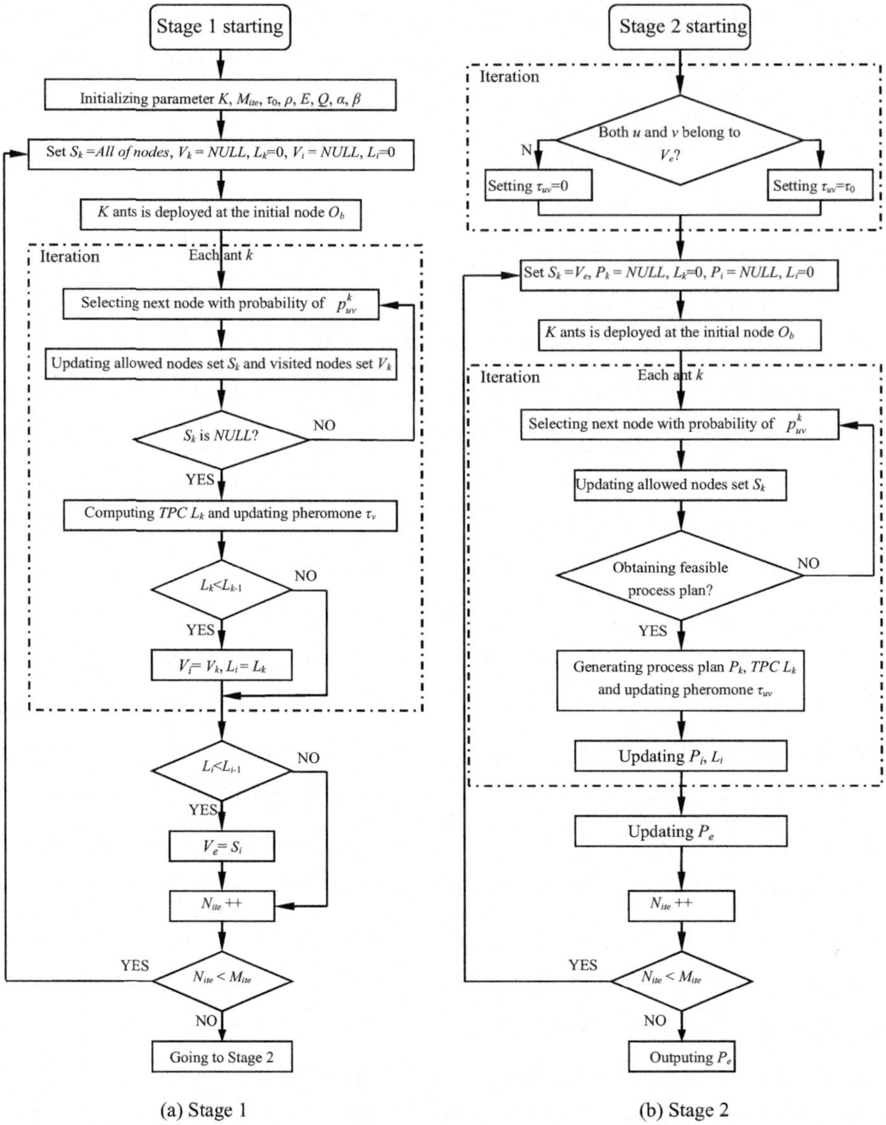

Fig. 3.4 Two-stage ACO approach

where E is a positive constant and PC is the processing cost of the selected node operation and it is calculated as follows.

$$PC = w_1 * MC + w_2 * TC \tag{3.20}$$

Table 3.8 Meaning of symbols

Symbol	Meaning
K	Number of ants
k	Index of ant, $k \in [1, K]$
u	Source node
v	Destination node
τ	Pheromone
η	Heuristic information
α	Relative weight of pheromone τ_{uv}
β	Relative weight of heuristic information η_{uv}
ρ	Pheromone evaporation rate
E	Algorithm constant to determine η_{uv}
Q	Algorithm constant to determine $\Delta\tau$
τ_0	Initial value of pheromone
S_k	Set of nodes allowed at the next step by ant k
V_k	Set of nodes visited by ant k
P_k	Process plan achieved by ant k
L_k	*TPC* achieved by ant k
P_i	Iteration-best process plan
V_i	Set of nodes generated by iteration-best process plan
L_i	Iteration-best *TPC*
V_e	The final set of nodes
P_e	The final process plan
M_{ite}	Maximum number of iterations
N_{ite}	Number of iteration

The Eq. (3.19) shows that the node with a smaller processing cost has a higher information value, that is to say it is more attractive to the ants.

The pheromone amount is initially set at τ_0 on every node. Pheromone intensity on the nodes is dynamically updated after ant colony has completed process plans. To avoid unlimited accumulation of the pheromone, the pheromone also evaporates at every round of iterations. The pheromone amount τ_v can be given as follows.

$$\tau_v = (1 - \rho) * \tau_v + \Delta\tau_v^k \tag{3.21}$$

where ρ is an evaporation coefficient of the pheromone on the destination node v. $\Delta\tau_v^k$ is the quantity of the pheromone increments on the node v generated by the ant k after each iteration. The amount of pheromone deposited on the node v by an ant k is proportional to a respective L_k. The process plans with smaller L_k will accumulate a greater amount of pheromone on their corresponding nodes. Also, can be given as:

$$\Delta\tau_v^k = \begin{cases} \frac{Q}{L_k}\ if\ ant\ k\ passes\ the\ node\ v \\ 0\ \ otherwise \end{cases} \tag{3.22}$$

where Q is a positive constant. L_k is the *TPC* of process plan generated by the ant k.

The heuristic information η_v and the pheromone amount τ_v constructed a probability of moving from a node from another node for an ant. The more the pheromone amount and the heuristic information on the nodes, the higher the selective probability. For the ant k, the selective probability p_v^k from the source node u to the destination node v can be calculated as following.

$$p_v^k = \begin{cases} \frac{[\tau_w]^\alpha[\eta_v]^\beta}{\sum_{w \in S_k}[\tau_w]^\alpha[\eta_w]^\beta} & v \in S_k \\ 0 & not\ v \in S_k \end{cases} \tag{3.23}$$

where α and β denote the weighting parameters controlling the relative importance of the pheromone amount and the heuristic information respectively.

In the initial state, when an ant k in the ant colony is located at the process node O_1 (OP_1, OPT_{11}), the set of selective node lists $G_1 = \{O_3, O_4, O_7, O_8, O_{14}, O_{15}\}$ in the next step. Assuming $w_1 = w_2 = 1, \alpha = \beta = 1$, calculate the selective probability of each node as shown in Table 3.9.

As shown in Table 3.9, O_4, O_8, and O_{15} have the same selective probability of 25.59%, which is higher than that of the other three nodes. The node O_8 of candidate operation Opt42 is selected according to the principle of random selection. The set of selective node lists $G_8 = \{O_3, O_4, O_9, O_{10}, O_{14}, O_{15}\}$ in the next step, calculate the selective probability is shown in Table 3.10.

Similarly, according to the principle of randomly selecting nodes under the same selective probability (27.79%), the node O_4 of candidate operation OPT_{22} is selected. The set of selective node lists $G_4 = \{O_5, O_6, O_9, O_{10}, O_{14}, O_{15}\}$ in the next step, calculate the selective probability is shown in Table 3.11.

The node O_{15} of candidate operation OPT_{82} with the maximum selective probability of 29.97% is selected. The set of selective node lists $G_{15} = \{O_5, O_6, O_9, O_{10}, O_{16}, O_{17}\}$ in the next step, calculate the selective probability is shown in Table 3.12.

Table 3.9 The selective probability of nodes for node O_1

Nodes	Operations	Candidate operations	Machine	Tool	TAD	Heuristic information η_{1v}	Selective probability p_{1v}^k (%)
O_3	OP_2	OPT_{21}	M1	T2	−Z	50/43 = 1.163	7.74
O_4		OPT_{22}	M2	T2	−Z	50/13 = 3.846	25.59
O_7	OP_4	OPT_{41}	M1	T4	−X	50/43 = 1.163	7.74
O_8		OPT_{42}	M2	T4	−X	50/13 = 3.846	25.59
O_{14}	OP_8	OPT_{81}	M1	T8	+X	50/43 = 1.163	7.74
O_{15}		OPT_{82}	M2	T8	+X	50/13 = 3.846	25.59

Table 3.10 The selective probability of nodes for node O_8

Nodes	Operations	Candidate operations	Machine	Tool	TAD	Heuristic information η_{8v}	Selective probability p_{8v}^k (%)
O_3	OP_2	OPT_{21}	M1	T2	− Z	50/43 = 1.163	8.40
O_4		OPT_{22}	M2	T2	− Z	50/13 = 3.846	27.79
O_9	OP_5	OPT_{51}	M1	T5	− X	50/48 = 1.042	7.53
O_{10}		OPT_{52}	M2	T5	− X	50/18 = 2.778	20.08
O_{14}	OP_8	OPT_{81}	M1	T8	+ X	50/43 = 1.163	8.40
O_{15}		OPT_{82}	M2	T8	+ X	50/13 = 3.846	27.79

Table 3.11 The selective probability of nodes for node O_4

Nodes	Operations	Candidate operations	Machine	Tool	TAD	Heuristic information η_{4v}	Selective probability p_{4v}^k (%)
O_5	OP_3	OPT_{31}	M1	T3	− Z	50/47 = 1.064	8.29
O_6		OPT_{32}	M2	T3	− Z	50/17 = 2.941	22.92
O_9	OP_5	OPT_{51}	M1	T5	− X	50/48 = 1.042	8.12
O_{10}		OPT_{52}	M2	T5	− X	50/18 = 2.778	21.65
O_{14}	OP_8	OPT_{81}	M1	T8	+ X	50/43 = 1.163	9.06
O_{15}		OPT_{82}	M2	T8	+ X	50/13 = 3.846	29.97

Table 3.12 The selective probability of nodes for node O_{15}

Nodes	Operations	Candidate operations	Machine	Tool	TAD	Heuristic information η_{15v}	Selective probability p_{15v}^k (%)
O_5	OP_3	OPT_{31}	M1	T3	− Z	50/47 = 1.064	9.14
O_6		OPT_{32}	M2	T3	− Z	50/17 = 2.941	25.26
O_9	OP_5	OPT_{51}	M1	T5	− X	50/48 = 1.042	8.95
O_{10}		OPT_{52}	M2	T5	− X	50/18 = 2.778	23.86
O_{16}	OP_9	OPT_{91}	M1	T9	+ X	50/48 = 1.042	8.95
O_{17}		OPT_{92}	M2	T9	+ X	50/18 = 2.778	23.86

Similarly, the node O_6 of candidate operation OPT_{32} with the maximum selective probability of 25.26% is selected. The set of selective node lists $G_6 = \{O_9, O_{10}, O_{16}, O_{17}\}$ in the next step, calculate the selective probability is shown in Table 3.13.

Similarly, according to the principle of randomly selecting nodes under the same selective probability (36.36%), the node O_{10} of candidate operation OPT_{52} is selected. The set of selective node lists $G_{10} = \{O_{11}, O_{16}, O_{17}\}$ in the next step, calculate the selective probability is shown in Table 3.14.

Table 3.13 The selective probability of nodes for node O_6

Nodes	Operations	Candidate operations	Machine	Tool	TAD	Heuristic information η_{6v}	Selective probability p_{6v}^k (%)
O_9	OP_5	OPT_{51}	M1	T5	− X	50/48 = 1.042	13.64
O_{10}		OPT_{52}	M2	T5	− X	50/18 = 2.778	36.36
O_{16}	OP_9	OPT_{91}	M1	T9	+ X	50/48 = 1.042	13.64
O_{17}		OPT_{92}	M2	T9	+ X	50/18 = 2.778	36.36

Table 3.14 The selective probability of nodes for node O_{10}

Nodes	Operations	Candidate operations	Machine	Tool	TAD	Heuristic information η_{10v}	Selective probability p_{10v}^k (%)
O_{11}	OP_6	OPT_{61}	M1	T6	+ Z	50/50 = 1	20.75
O_{16}	OP_9	OPT_{91}	M1	T9	+ X	50/48 = 1.042	21.62
O_{17}		OPT_{92}	M2	T9	+ X	50/18 = 2.778	57.63

Similarly, the node O_{17} of candidate operation OPT_{92} with the maximum selective probability of 57.63% is selected. So, the node O_{11} of candidate operation OPT_{61} and the node O_{12} of candidate operation OPT_{71} are selected.

Figure 3.3 can be simplified after Stage 1, as shown in Fig. 3.5.

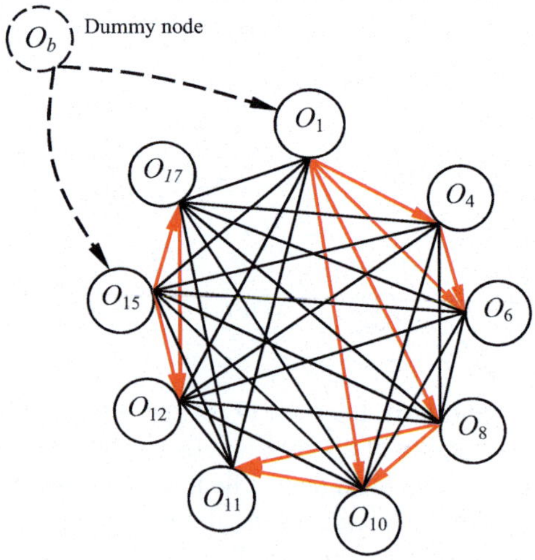

Fig. 3.5 Simple graph with reduced nodes after Stage 1

2. **Stage 2—Operation sequencing**

If the set of node V_e is decided after Stage 1, the ants will be only allowed to travel among the nodes. Those nodes which have not been selected in Stage 1 will be ignored in Stage 2. In contrast to Stage 1, arcs of the weight graph are the pheromone carriers in Stage 2. All the pheromone values on arcs (u, v) that involve those unselected processes would be set to 0.

For the Part 1 in Fig. 3.2, after Stage 1, the graph in Fig. 3.3 is effectively reduced to a simple graph depicted in Fig. 3.5 which contains only the nodes included in set V_e. The ants have to visit all nodes on the simple graph along the arcs connected each pair of nodes. Thus, the ants deposit pheromones on the arcs of the graph. The transition probability from node u to v for the ant k is given as:

$$p_v^k = \begin{cases} \frac{[\tau_w]^\alpha [\eta_v]^\beta}{\sum_{w \in S_k} [\tau_w]^\alpha [\eta_w]^\beta} & v \in S_k \\ 0 & not\ v \in S_k \end{cases} \tag{3.24}$$

The heuristic information η_v is calculated as the Eqs. (3.19) and (3.20). The pheromone amount τ_{uv} on the arcs (u, v) can be given as follows.

$$\tau_v = (1-\rho) * \tau_v + \Delta\tau_v^k \tag{3.25}$$

$$\Delta\tau_v^k = \begin{cases} \frac{Q}{L_k}\ if\ ant\ k\ passes\ the\ node\ v \\ 0\ \ otherwise \end{cases} \tag{3.26}$$

In Stage 2, ant colony travel through the nodes and arcs that are still remaining, as in Fig. 3.5. The pheromone account on the arcs and the heuristic information on the nodes will guide ants to visit all of nodes. The final visited path generated by ant colony represents the process planning result with minimal *TPC*. Based on the simple graph in Fig. 3.5, the final visited path is show in Fig. 3.6. The corresponding process plan and its cost can be shown in Table 3.15.

When Q is 2000, the pheromone increment on the corresponding path of the above process plan is shown in Eq. (3.27).

$$\Delta\tau_{uv}^k = \frac{Q}{L_k} = \frac{2000}{1514} = 1.321 \tag{3.27}$$

When $\tau_0 = 1$ and $\rho = 0.75$, the pheromone on the corresponding path is shown in Eq. (3.28).

$$\tau_{uv} = (1 - 0.75) * 1 + 1.321 = 1.571 \tag{3.28}$$

For Part 1, 5 trials were separately conducted to evaluate the performance of the two-stage ACO approach. Experimental observation has shown that $K = 10$, $\rho = 0.75$, $\alpha = 1$, $\beta = 1$, $E = 50$, $Q = 2000$, $\tau_0 = 1$, $M_{ite} = 50$ are the choices of these parameters. The final visited path is shown in Fig. 3.7, and the corresponding process plans is listed in Table 3.16.

3.4.3 Case

Two prismatic parts are used for the case experiments. The first prismatic part (Part 2) used by Zhang et al. [1] is illustrated in Fig. 3.8. It consists of 14 STEP-defined manufacturing features and 14 machining operations. The machining information and precedence constraints are given in Tables 3.17, and 3.18. The second prismatic part (Part 3) used by Li et al. [2] is illustrated in Fig. 3.9. The machining information and precedence constraints are given in Tables 3.19 and 3.20.

3.4.4 Experiments and Results

Two conditions are used to test proposed algorithm on Part 2 in Fig. 3.8

(1) All machines and tools are available, and w_1-w_6 in Eqs. (3.18) and (3.20) are set to 1.
(2) All machines and tools are available, and

$$w_2 = w_5 = 0, w_1 = w_3 = w_4 = w_6 = 1 \tag{3.29}$$

For Part 2, 20 trials were separately conducted to evaluate the performance of the two-stage ACO approach under condition (1) and (2) respectively. Experimental

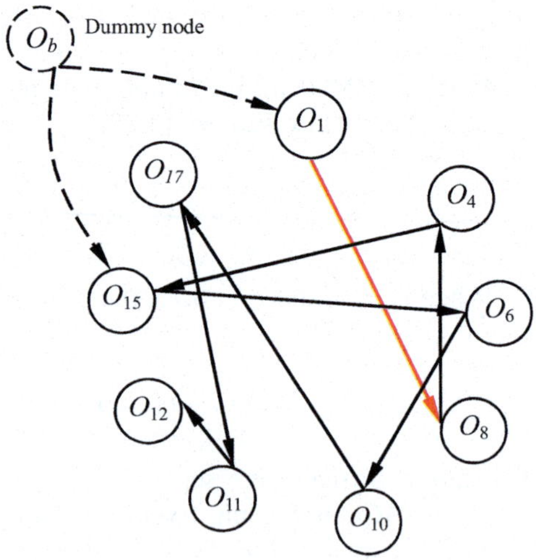

Fig. 3.6 Final visited path after Stage 2

Table 3.15 The final process plan and its cost

Node	Operations	Candidate operations	Machine	Tool	TAD	*MC*	*TC*	*MCC*	*TCC*	*SCC*
O_1	OP_1	OPT_{11}	M1	T1	+ X	40	10			
O_8	OP_4	OPT_{42}	M2	T4	− X	10	3	300	60	20
O_4	OP_2	OPT_{22}	M2	T2	− Z	10	3		60	20
O_{15}	OP_8	OPT_{82}	M2	T8	+ X	13	7		60	20
O_6	OP_3	OPT_{32}	M2	T3	− Z	10	7		60	20
O_{10}	OP_5	OPT_{52}	M2	T5	− X	10	8		60	20
O_{17}	OP_9	OPT_{92}	M2	T9	+ X	10	10		60	20
O_{11}	OP_6	OPT_{61}	M1	T6	+ Z	50	3	300	60	20
O_{12}	OP_7	OPT_{71}	M1	T7	− Z	40	10		60	20
Subtotal						193	61	600	480	180
Subtotal						254		1260		
Total						1514				

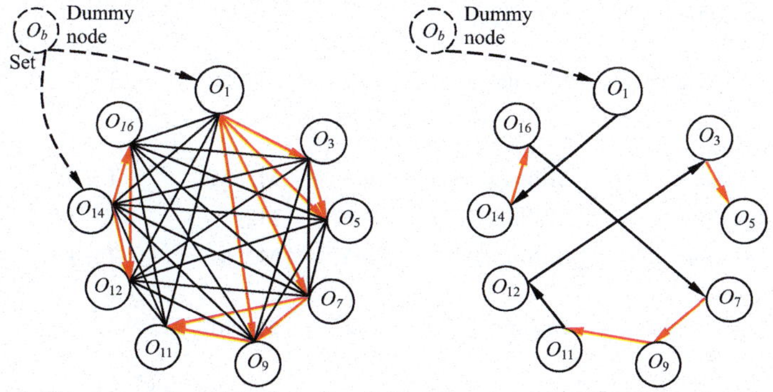

(a) Simple graph with reduced nodes after Stage 1

(b) Final visited path

Fig. 3.7 Disjunctive graph and ant visited path

Table 3.16 Optimal process plans for the Part 1

Node	O_1	O_{14}	O_{16}	O_7	O_9	O_{11}	O_{12}	O_3	O_5
Operation	OP_1	OP_8	OP_9	OP_4	OP_5	OP_6	OP_7	OP_2	OP_3
Machine	M1	M1	M1	M1	M1	M1	M1	M1	M1
Tool	T1	T8	T9	T4	T5	T6	T7	T2	T3
TAD	+ X	+ X	+ X	− X	− X	+ Z	− Z	− Z	− Z

$NMC = 0$, $NCC = 8$, $NSC = 3$. $TMC = 360$, $TTC = 62$, $TMCC = 0$, $TTCC = 160$, $TSCC = 240$, $TPC = 822$

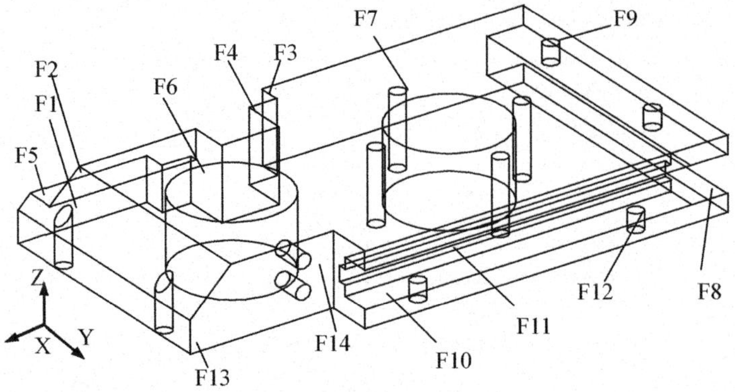

Fig. 3.8 A sample part with 14 features and 14 operations—Part 2

Table 3.17 Features, operations and machining information for Part 2

Features	Feature descriptions	Operations	TADs	Machines	Tools	Remarks
F1	Two replicated holes	Drilling(OP_1)	+ Z, − Z	M1, M2, M3	T1	M1(10): drill press M2(35): vertical milling M3(60): vertical CNC milling T1(3): drill 1 T2(3): drill 2 T3(8): reamer T4(15): boring tool T5(10): milling cutter 1 T6(15): milling cutter 2 T7(10): chamfer tool T8(10): slot cutter MCC = 300 SCC = 120 TCC = 15
F2	A chamfer	Milling(OP_2)	− X, + Y, − Y, − Z	M2, M3	T8	
F3	A slot	Milling(OP_3)	+ Y	M2, M3	T5, T6	
F4	A slot	Milling(OP_4)	+ Y	M2	T5, T6	
F5	A step	Milling(OP_5)	+ Y, − Z	M2, M3	T5, T6	
F6	Two replicated holes	Drilling(OP_6)	+ Z, − Z	M1, M2, M3	T2	
F7	Four replicated holes	Drilling(OP_7)	+ Z, − Z	M1, M2, M3	T1	
F8	A slot	Milling(OP_8)	+ X	M2, M3	T5, T6	
F9	Two replicated holes	Drilling(OP_9)	− Z	M1, M2, M3	T1	
F10	A slot	Milling(OP_{10})	− Y	M2, M3	T5, T6	
F11	A slot	Milling(OP_{11})	− Y	M2, M3	T5, T7	
F12	Two replicated holes	Drilling(OP_{12})	+ Z, − Z	M1, M2, M3	T1	
F13	A step	Milling(OP_{13})	− X, − Y	M2, M3	T5, T6	
F14	Two replicated holes	Drilling(OP_{14})	− Y	M1, M2, M3	T1	

Table 3.18 Precedence constraints for Part 2

Constraints	Descriptions	Hard or soft
Tool interactions	OP_1 should be prior to OP_2	Hard
Datum interactions	OP_6 should be prior to OP_7 OP_{10} should be prior to OP_{11} OP_{13} should be prior to OP_{14}	Hard
Thin-wall interactions	OP_9 should be prior to OP_8 OP_{12} should be prior to OP_{10}	Soft
Material removal interactions	OP_8 should be prior to OP_9 OP_{10} should be prior to OP_{12} OP_{13} should be prior to OP_{14} OP_3 should be prior to OP_4	Soft

observation has shown that $K = 25$, $\rho = 0.75$, $\alpha = 1$, $\beta = 1$, $E = 50$, $Q = 2000$, $\tau_0 = 1$, $M_{ite} = 200$ are the best choices of these parameters. The best process plans generated respectively under condition (1) and (2) is listed in Table 3.21.

The comparing results with those of GA, SA approach by Li et al. [1], TS by Li et al. [3], as well as the ACO by Liu et al. [17] are shown in Table 3.22.

Under condition (1), among 20 trial results, *TPC* (1328.0) occurs 18 times, *TPC* (1348) occurs 1 times, and *TPC* (1343) occurs 1 times. The mean *TPC* (1329.0) is better than the costs of other four algorithms. Under condition (2), *TPC* (1170.0) occurs 20 times in 20 trials, which is better than the performance of TS [3], SA and GA [1] and is the same as ACO [17].

Besides the above two conditions, an additional condition are used to test the two-stage ACO approach on Part 3 in Fig. 3.9 as follows:

(3) Machine M2 and tool T7 are down, and

$$w_2 = w_5 = 0, w_1 = w_3 = w_4 = w_6 = 1 \tag{3.30}$$

For Part 3, 20 trials were separately conducted to evaluate the performance of the proposed approach under condition (1), (2) and (3) respectively. Experimental observation has shown that $K = 40$, $\rho = 0.75$, $\alpha = 1$, $\beta = 1$, $E = 100$, $Q = 3000$, $\tau_0 = 1$, $M_{ite} = 200$ are the best choices of these parameters. The best process plans generated respectively under condition (1), (2) and (3) is listed in Table 3.23. The comparing results with those of GA, SA approach by Li et al. [1], TS by Li et al. [3], as well as the HGA by Huang et al. [6] are shown in Table 3.24.

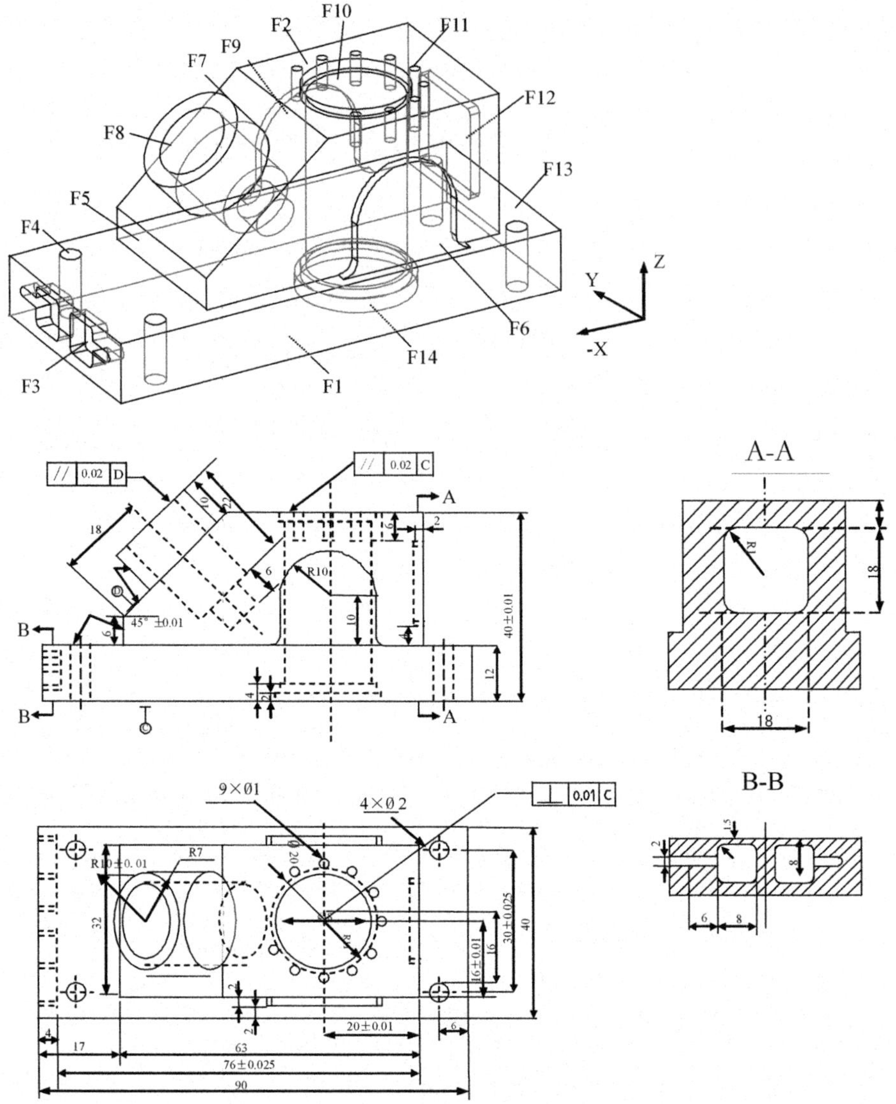

Fig. 3.9 A sample part with 14 features and 20 operations—Part 3

Under condition (1), the better performance of the two-stage approach with minimum *TPC* (2525.0) has been found. In addition, the all costs including mean *TPC* (2552.4), maximum *TPC* (2557), and minimum *TPC* (2525.0) are better than the costs achieved by other algorithms. Under conditions (2), a lower *TPC* (2090.0) has been found using the two-stage ACO approach. The maximum *TPC* (2380.0) is the same as the result of SA and is superior to the result of TS and GA. Under condition (3), the minimum *TPC* (2590) is inferior to the result of TS and is the same

Table 3.19 Features, operations and machining information for Part 3

Features	Feature descriptions	Operations	TADs	Machines	Tools	Remarks
F1	Planar surface	Milling (OP_1)	+ Z	M2, M3	T6, T7, T8	M1(10): drilling press M2(40): three-axis vertical milling machine M3(100): CNC 3-axis vertical milling machine M4(60): boring machine T1(7): drill 1 T2(5): drill 2 T3(3): drill 3 T4(8): drill 4 T5(7): tapping tool T6(10): mill 1 T7(15): mill 2 T8(30): mill 3 T9(15): ream T10(20): boring tool MCC = 160 SCC = 100 TCC = 20
F2	Planar surface	Milling (OP_2)	− Z	M2, M3	T6, T7, T8	
F3	Two replicated pockets	Milling (OP_3)	+ X	M2, M3	T6, T7, T8	
F4	Four replicated holes	Drilling (OP_4)	+ Z, − Z	M1, M2, M3	T2	
F5	A step	Milling (OP_5)	+ X, − Z	M2, M3	T6,T7	
F6	A protrusion (rib)	Milling (OP_6)	+ Y, − Z	M2, M3	T7, T8	
F7	A boss	Milling (OP_7)	− a	M2, M3	T7, T8	
F8	A compound hole	Drilling (OP_8)	− a	M1, M2, M3	T2, T3, T4	
		reaming (OP_9)		M1, M2, M3	T9	
		Boring (OP_{10})		M3, M4	T10	
F9	A protrusion (rib)	Milling (OP_{11})	− Y, − Z	M2, M3	T7, T8	
F10	A compound hole	Drilling (OP_{12})	− Z	M1, M2, M3	T2, T3, T4	
		Reaming (OP_{13})		M1, M2, M3	T9	
		Boring (OP_{14})		M3, M4	T10	
F11	Nine replicated holes	Drilling (OP_{15})	-Z	M1, M2, M3	T1	
		Tapping (OP_{16})		M1, M2, M3	T5	
F12	A pocket	Milling (OP_{17})	− X	M2, M3	T7, T8	
F13	A step	Milling (OP_{18})	− X, − Z	M2, M3	T6, T7	
F14	A compound hole	Reaming (OP_{19})	+ Z	M1, M2, M3	T9	
		Boring (OP_{20})		M3, M4	T10	

Table 3.20 Precedence constraints for Part 3

Features	Operation	Precedence constraints description	Hard or soft
F1	Milling (OP_1)	F1 (OP_1) is the datum and supporting face for the part, hence it is machined prior to all features and operations	Hard
F2	Milling (OP_2)	F2 (OP_1) is prior to F10 (OP_{12}, OP_{13}, OP_{14}) and F11 (OP_{15}, OP_{16}) for the material removal interactions	Hard
F5	Milling (OP_5)	F5(OP_5) is prior to F4 (OP_4) and F7(OP_7) for the datum interactions	Hard
F6	Milling (OP_6)	F6 (OP_6) is prior to F10 (OP_{12}, OP_{13}, OP_{14}) for the datum interaction	Hard
F7	Milling (OP_7)	F7 (OP_7) is prior to F8 (OP_8, OP_9, OP_{10}) for the datum interactions	Hard
F8	Drilling (OP_8) Reaming (OP_9) Boring (OP_{10})	OP_8 is prior to OP_9 and OP_{10}, OP_9 is prior to OP_{10} for the fixed order of machining operations	Hard
F9	Milling (OP_{11})	F9 (OP_{11}) is prior to F10 (OP_{12}, OP_{13}, OP_{14}) for the datum interaction	Hard
F10	Drilling (OP_{12}) Reaming (OP_{13}) Boring (OP_{14})	OP_{12} is prior to OP_{13} and OP_{14}, OP_{13} is prior to OP_{14} for the fixed order of machining operations. F10 (OP_{12}, OP_{13}, OP_{14}) is prior to F11 (OP_{15}, OP_{16}), and OP_{12} of F10 is prior to F14 (OP_{19}, OP_{20}) for the datum interaction	Hard
F11	Drilling (OP_{15}) Tapping (OP_{16})	OP_{15} is prior to OP_{16} for the fixed order of operations	Hard
F13	Milling (OP_{18})	F13 (OP_{18}) is prior to F4 (OP_4) and F12 (OP_{17}) for the material removal interaction	Soft
F14	Reaming (OP_{19}) Boring (OP_{20})	OP_{19} is prior to OP_{20} for the fixed order of machining operations	Hard

as the result of SA. The mean *TPC* (2600.8) is the better than the result of TS, SA and GA. In general, the performance of proposed algorithm for Part 3 is better than the other four algorithms.

3.5 Solving Process Planning Problem Using a Particle Swarm Optimization

3.5.1 Particle Swarm Optimization Approach

PSO is a swarm optimization approach [16]. Every particle in the population represents an *N*-dimensional solution that constructs a search space for every particle.

Table 3.21 The best process plans for Part 2 corresponding to condition (1) and (2)

Condition (1)														
Operation	6	1	7	9	12	5	3	4	8	10	11	13	14	2
Machine	2	2	2	2	2	2	2	2	2	2	2	2	2	2
Tool	2	1	1	1	1	5	5	5	5	5	5	5	1	8
TAD	− Z	− Z	− Z	− Z	− Z	− Z	+ Y	+ Y	+ X	− Y	− Y	− Y	− Y	− Y
$NMC = 0$, $NTC = 4$, $NSC = 3$, $NPC = 1$, $TMCC = 0$, $TTCC = 60$, $TSCC = 480$, $TMC = 490$, $TTC = 98$, $TAPC = 200$, $TPC = 1328$														
Condition (2)														
Operation	6	7	12	1	9	2	5	13	8	3	4	10	11	14
Machine	2	2	2	2	2	2	2	2	2	2	2	2	2	2
Tool	2	1	1	1	1	8	5	5	5	5	5	5	5	1
TAD	− Z	− Z	− Z	− Z	− Z	− Z	− Z	− Z	+ X	+ Y	+ Y	− Y	− Y	− Y
$NMC = 0$, $NSC = 3$, $NPC = 1$, $TMCC = 0$, $TSCC = 480$, $TMC = 490$, $TAPC = 200$, $TPC = 1170$														

Table 3.22 Results compared to other algorithm for Part 1

Condition	Proposed approach	ACO	TS	SA	GA
(1)					
Mean	1329.0	1329.5	1342.6	1373.5	1611.0
Maximum	1348.0	1343.0	1378.0	1518.0	1778.0
Minimum	1328.0	1328.0	1328.0	1328.0	1478.0
(2)					
Mean	1170.0	1170.0	1194.0	1217.0	1482.0
Maximum	1170.0	1170.0	1290.0	1345.0	1650.0
Minimum	1170.0	1170.0	1170.0	1170.0	1410.0

The particles fly freely to search for the optimal position at a given velocity. Hence, for the particle i, the vectors X_i^t and V_i^t at the tth iteration can be denoted as $X_i^t = \{X_{i1}^t, X_{i2}^t, \ldots, X_{ik}^t, \ldots, X_{iN}^t\}$ and $V_i^t = \{V_{i1}^t, V_{i2}^t, \ldots, V_{ik}^t, \ldots, V_{iN}^t\}$, respectively. With the emergence of the optimal position at iteration $t+1$, the vectors X_i^t and V_i^t can be updated as follows:

$$V_{ik}^{t+1} = w * V_{ik}^t + c_1 * Rand() * \left(P_{ik}^t - X_{ik}^t\right) + c_2 * Rand() * \left(P_{gk}^t - X_{ik}^t\right) \tag{3.31}$$

$$X_{ik}^{t+1} = X_{ik}^t + V_{ik}^{t+1} \tag{3.32}$$

In Eqs. (3.31) and (3.32), V_{ik}^{t+1} is the velocity on dimension k, and X_{ik}^{t+1} is the position on dimension k. P_{ik}^t is the local optimal position on dimension k, and P_{gk}^t is the global optimal position on dimension k. The weight w is used to control the iteration. The constants c_1 and c_2 control the balance between a local optimal position and the global optimal position. Rand () is limited in [0, 1]. The traditional PSO approach is carried out in four steps:

Step 1: Population initialization.
Step 2: Update the vectors X_i^t and V_i^t according to Eqs. (3.31) and (3.32).
Step 3: Update P_{ik}^t and P_{gk}^t according to the performance of the population.
Step 4: Loop to Step 2 until a termination condition is met.

3.5.2 Solution Representation

It is necessary to modify the traditional PSO for the PP problem to include, for example, particle representation and the particle movement strategy. In applying the PSO approach to the PP problem, three tasks have to be performed, namely, particle encoding, particle validation and particle decoding. The first task is to encode a

Table 3.23 The best process plans corresponding to condition (1), (2) and (3)

Condition (1)																				
Operation	1	3	5	6	2	18	11	12	13	17	7	8	9	19	14	20	10	4	15	16
Machine	2	2	2	2	2	2	2	2	2	2	2	2	2	2	4	4	4	1	1	1
Tool	6	6	6	6	6	6	7	3	9	7	7	2	9	9	10	10	10	2	1	5
TAD	+ Z	+ X	+ X	− Z	− Z	− Z	− Z	− Z	− Z	− X	− a	− a	− a	+ Z	− Z	+ Z	− a	− Z	− Z	− Z
NMC = 2, *NTC* = 10, *NSC* = 10, *TMCC* = 320, *TTCC* = 200, *TSCC* = 1000, *TMC* = 770, *TTC* = 235 *TPC* = 2525																				
Condition (2)																				
Operation	1	2	18	11	6	12	13	19	17	3	5	7	8	9	10	20	14	4	15	16
Machine	2	2	2	2	2	2	2	2	2	2	2	2	2	2	4	4	4	1	1	1
Tool	7	7	7	7	7	3	9	9	7	7	7	7	3	9	10	10	10	2	1	5
TAD	+ Z	− Z	− Z	− Z	− Z	− Z	− Z	+ Z	− X	+ X	+ X	− a	− a	− a	− a	+ Z	− Z	− Z	− Z	− Z
NMC = 2, *NSC* = 8, *TMCC* = 320, *TSCC* = 1000, *TMC* = 770, *TPC* = 2090																				
Condition (3)																				
Operation	1	6	2	5	11	12	13	14	18	17	7	8	9	10	19	20	3	4	15	16
Machine	3	3	3	3	3	3	3	3	3	3	3	3	3	3	3	3	3	1	1	1
Tool	6	6	6	6	8	2	9	10	6	8	8	2	9	10	9	10	6	2	1	5
TAD	+ Z	− Z	− Z	− Z	− Z	− Z	− Z	− Z	− X	− X	− a	− a	− a	− a	+ Z	+ Z	+ X	− Z	− Z	− Z
NMC = 1, *NSC* = 6, *TMCC* = 160, *TSCC* = 700, *TMC* = 1730, *TPC* = 2590																				

Table 3.24 Results compared to other algorithm for Part 3

Condition	Proposed approach	HGA	TS	SA	GA
(1)					
Mean	2552.4	–	2609.6	2668.5	2796.0
Maximum	2557.0	–	2690.0	2829.0	2885.0
Minimum	2525.0	2527.0	2527.0	2535.0	2667.0
(2)					
Mean	2120.5	–	2208.0	2287.0	2370.0
Maximum	2380.0	–	2390.0	2380.0	2580.0
Minimum	2090.0	2120.0	2120.0	2120.0	2220.0
(3)					
Mean	2600.8	–	2630.0	2630.0	2705.0
Maximum	2740.0	–	2740.0	2740.0	2840.0
Minimum	2590.0	–	2580.0	2590.0	2600.0

Table 3.25 Detail of a particle

Data type	Variable	Description
Integer	*ix_o*	Id of an operation, which corresponds to the index of the operation set
Integer	*ix_m*	Id of a machine, which corresponds to the index of the machine set
Integer	*ix_c*	Id of a cutting tool, which corresponds to the index of the cutting tool set
Integer	*ix_t*	Id of a TAD, which corresponds to the index of the TAD set

process plan to an appropriate particle. Because operation determining and operation sequencing have to be performed simultaneously in process planning, the particle structure should be considered to comprise the information of determining and sequencing operations. The details of a particle are listed in Table 3.25.

According to the definition of a particle, we modified a particle i of a $2 \times n$ matrix to represent a process plan at iteration t [11], i.e.,

$$X_i^t = \begin{bmatrix} x_{i11}^t, x_{i12}^t, \ldots, x_{i1n}^t \\ x_{i21}^t, x_{i22}^t, \ldots, x_{i2n}^t \end{bmatrix} \tag{3.33}$$

The first row x_{i1n}^t is the Operation Determining (OD) part and represents the operation selection for each feature. The second row x_{i2n}^t is the Operation Sequencing (OS) part and represents the priority among operations. x_{i2n}^t is initialized randomly in the range of 0 and 1 according to the priority among operations, and x_{i1n}^t can be calculated by Eq. (3.34).

$$X_{i1n}^t = \left(a^2 * ix\_m + a * ix\_c + ix\_t\right)/a^3 \tag{3.34}$$

In Eq. (3.34), *ix_m*, *ix_c*, and *ix_t* are generated randomly from the candidate machine set, cutting tool set, and TAD set for executing the operation. a can be given by the equation.

$$a = Max(N_M, N_T, N_S) + 1 \tag{3.35}$$

where N_S is the number of TADs.

To illustrate the particle encoding, the process plan of Part 1 in Table 3.26 is taken as an example and the corresponding encoding of x_{i1n}^t is shown in Fig. 3.10. For instance, the values of *ix_m*, *ix_*c, and *ix_t* are set to 1 if the OPTs of M2, T1 and + X, respectively, are selected to process operation OP_1. If the OPTs of M2, T1 and + Z are selected to process operation OP_1, the values of *ix_m*, *ix_c*, and *ix_t* will be 1, 1 and 3, respectively. For the process plan in Table 3.26, a feasible particle is listed in Table 3.27.

In Table 3.27, the first row means the operations. The second row represents the calculated value of its assigned OPT, and the third row represents the priority among the operations.

Table 3.26 Available process plan for Part 1

Operation	Candidate operations	Machine	Tool	TAD
OP_1	OPT_{11}	M1	T1	+ X
OP_2	OPT_{22}	M2	T2	− Z
OP_3	OPT_{31}	M1	T3	− Z
OP_4	OPT_{41}	M1	T4	− X
OP_5	OPT_{52}	M2	T5	− X
OP_6	OPT_{61}	M1	T6	+ Z
OP_8	OPT_{81}	M1	T8	+ X
OP_9	OPT_{92}	M2	T9	+ X
OP_7	OPT_{71}	M1	T7	− Z

Operation	OP_1	OP_8	OP_9	OP_4	OP_5	OP_6	OP_7	OP_2	OP_3
x_{i1n}^t	0.111	0.181	0.191	0.144	0.154	0.163	0.176	0.126	0.136
(*ix_m*, *ix_c*, *ix_t*)	(1, 1, 1)	(1, 8, 1)	(1, 9, 1)	(1, 4, 4)	(1, 5, 4)	(1, 6, 3)	(1, 7, 6)	(1, 2,6)	(1, 3, 6)

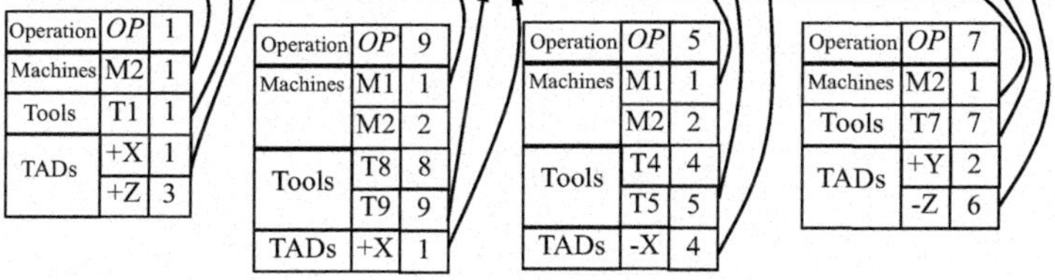

Fig. 3.10 Encoding of x_{i1n}^t for the process plan in Table 3.26

Table 3.27 A feasible particle for the process plan in Table 3.26

Operation	OP_1	OP_2	OP_3	OP_4	OP_5	OP_6	OP_7	OP_8	OP_9
x^t_{i1n}	0.111	0.126	0.136	0.144	0.154	0.164	0.176	0.181	0.191
x^t_{i2n}	1	0.50	0.40	0.85	0.75	0.70	0.60	0.95	0.90

The second task is to validate the particles to accord with the precedence constraints. A particle represents a process plan. Nevertheless, process plans generated by particles flying freely in solution space are usually invalid against the precedence constraints. To validate each particle, an $n \times n$ constraint matrix P is proposed to incorporate the precedence constraints among the operations into the particles.

$$P = \left(p_{ij}\right)n \times n \tag{3.36}$$

$$P_{ij} = \begin{cases} 1 \; OP_i \; is \; prior \; to \; OP_j \; or \; i = j \\ 0 \; otherwise \end{cases} \tag{3.37}$$

The precedence constraint matrix for the precedence constraints in Table 3.3 is shown as follows.

$$\begin{bmatrix} 1\,0\,0\,0\,0\,0\,0\,0\,0 \\ 1\,1\,0\,0\,0\,1\,0\,0\,0 \\ 1\,1\,1\,0\,0\,1\,0\,0\,0 \\ 1\,0\,0\,1\,0\,0\,0\,0\,0 \\ 1\,0\,0\,1\,1\,0\,0\,0\,0 \\ 0\,0\,0\,1\,1\,1\,0\,0\,0 \\ 0\,0\,0\,0\,0\,0\,1\,1\,1 \\ 0\,0\,0\,0\,0\,0\,0\,1\,0 \\ 0\,0\,0\,0\,0\,0\,0\,1\,1 \end{bmatrix} \tag{3.38}$$

The third task is to decode a particle into a solution. According to the input of encoded position matrix, a particle can be decoded to obtain a process plan. The particle decoding includes the following two steps.

First, determine the machine, cutting tool and TAD according to the value x^t_{i1n}. The decoding approach for x^t_{i1n} is

$$y_{in} = x^t_{i1n} * a^3 \tag{3.39}$$

$$ix\_m_{in} = y_{in}/a^2 \tag{3.40}$$

$$ix\_c_{in} = \left(y_{in} - ix\_m_{in} * a^2\right)/a \tag{3.41}$$

$$ix\_t_{in} = y_{in} - ix\_m_{in} * a^2 - ix\_c_{in} * a \tag{3.42}$$

where y_{in} is an integer.

Second, sequence these determined operations according to the precedence value x_{i2n}^t. If the sequence of operations violates the precedence constraints, the operations should be sequenced again according to the precedence value x_{i2n}^t with the help of the precedence matrix P_m.

$$x_{i2n}^t = x_{i2n}^t \times P_m \tag{3.43}$$

Similarly, for particle *i*, a velocity matrix at iteration *t* can be represented as

$$V_i^t = \left[v_{i1}^t, v_{i2}^t, \ldots, v_{ik}^t, \ldots, v_{in}^t\right] \tag{3.44}$$

where v_{in}^t is initialized randomly in the range of -1 to 1.

3.5.3 *Particles Operations*

1. **Population initialization**

The particle swarm is initialized in three steps.

(i) Set the population size P_{max} and the maximum iteration number N_{max}.
(ii) Initialize each particle. The initial position and velocity of each of the particles in the population are generated.
(iii) Decode every particle to the process plan according to Eqs. (3.39)–(3.43), and then calculate *CP*. Get the local optimal position P_i^0 and the global optimal position P_g^0.

2. **Iteration and control**

For every selected particle, the position and velocity can be updated according to Eqs. (3.31) and (3.32). To ensure the process plan validity, decoding every newly generated particle to the process plan is necessary. If the new process plans violate the precedence constraints, the corresponding particles will be normalized by using the constraint matrix *P*. For each valid particle, the *CP* of the corresponding process plan will be calculated. If a lower *CP* is achieved, the local optimal position P_l and the global optimal position P_g will be updated.

When the traditional PSO is applied in the PP problem, a quick local convergence at an early stage of the PSO usually has to be faced. The quick local convergence will make further exploration difficult and can generate an undesirable solution. To solve this problem, some modifications are suggested to enhance the performance of the traditional PSO algorithm [9, 10, 13]. Four operators are incorporated into the traditional PSO approach. Because the position of a particle is expressed as a $2 \times n$

matrix, in which the first row is the *OD* part and the second row is the *OS* part, the operators for the different rows vary independently.

For the *OD* part, two types of mutation operator are designed to generate a new solution.

(1) Mutation operator 1

One particle in the swarm is chosen for a mutation operation with a predefined probability (p_{ms}). First, for the *OD* part of this particle, one position point L is randomly selected. Second, decode the particle to a process plan, and obtain the machine, cutting tool, and TAD (M, T, TAD) of the Lth operation in this process plan. Third, from the machine set, cutting tool set, and TAD set of the Lth operation, an alternative selection of machine, cutting tool, and TAD is determined to replace the current machine, cutting tool, and TAD.

(2) Mutation operator 2

One particle is chosen for a mutation operation with a predefined probability (p_{ss}). For the *OD* part of this particle, two adjacent position points L_1 and L_2 are randomly selected. Decode the particle to a process plan, and obtain the machines, cutting tools, and TADs (Ms, Ts, TADs) of the L_1th and L_2th operations in this process plan. If $M_{L_2} = M_{L_1}$, let $T_{L_1} = T_{L_2}$ or $TAD_{L_1} = TAD_{L_2}$, or otherwise the position points L_1, L_2 will be reselected.

With respect to the *OS* part, two operators are employed, crossover and shift.

(1) Crossover operator
Two particles, A and B, are selected to execute a crossover operation with a predefined probability (p_{cq}). For the *OS* part of those two particles, one position point L is randomly determined. The *OS* part is divided into two parts. Subsequently, the value of the front part of OS_A is taken out and inserted before the cutting point of OS_B, and the value of the left part of OS_B is taken out and inserted before the cutting point of OS_A.
(2) Shift operator
One particle is chosen for a shift operation with a predefined probability (p_{sq}). For the *OS* part of this particle, two position points L_1, L_2 are randomly selected, and their values $x_{i1L_1}^t$, $x_{i1L_2}^t$ exchanged.
(3) **Termination**
If the max number of iterations N_{max} is reached, the iteration will be terminated. Decode the obtained particle position P_g^t to achieve the final process plan.

The flowchart is described in Fig. 3.11.

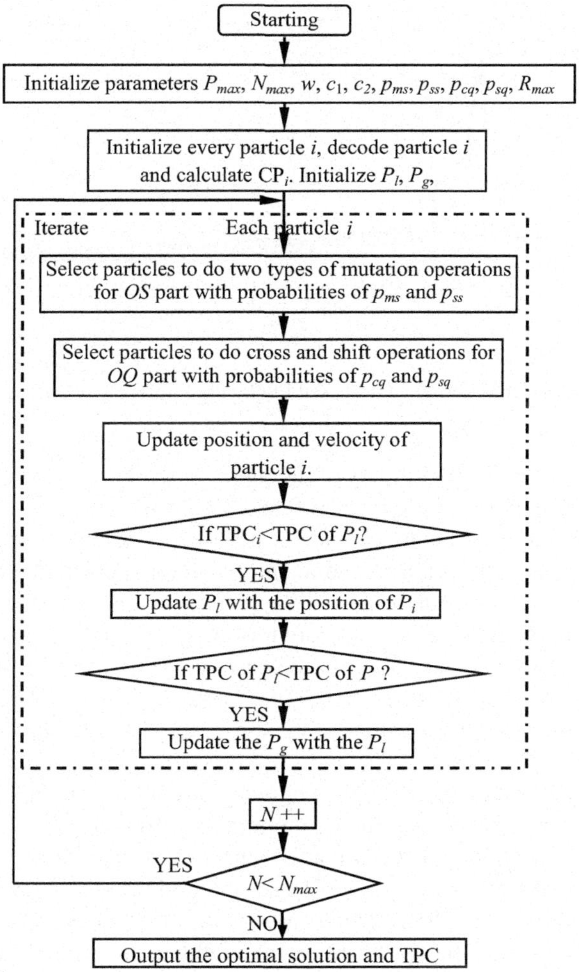

Fig. 3.11 Flowchart of the modified PSO approach

3.5.4 Experiments and Results

While applying PSO to solve the process planning problem, the key parameters have to be determined to facilitate the performance of the modified PSO. Accordingly, many preliminary simulation experiments have to be carried out to determine those parameters. The process planning problem for Part 2 in Sect. 3.4.3 is used to illustrate how the key parameters are determined. It is assumed that w_1–w_5 in Eq. (3.18) are set as 1.

The key parameters may be analysed from two aspects, namely, the swarm characteristic parameters of PSO (P_{max}, N_{max}, w, c_1, c_2) and the problem data (p_{ms}, p_{ss}, p_{cq}, p_{sq}). The swarm size P_{max} and iteration number N_{max} will be increased with the

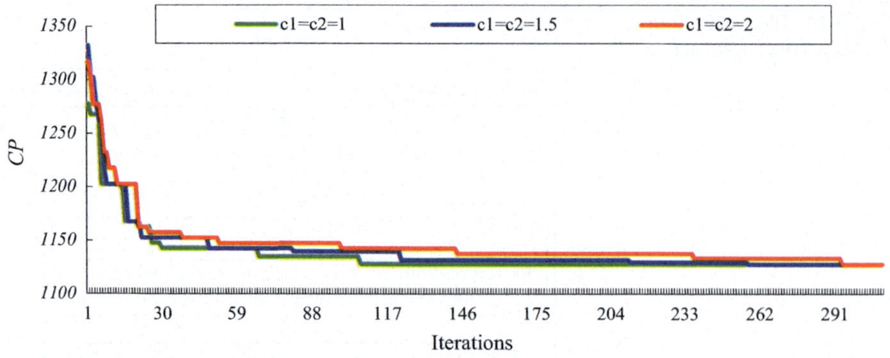

Fig. 3.12 *CP* of the proposed PSO with different constants c_1 and c_2

increased complexity of the part. For the Part 2 in Sect. 3.4.3, after many trials, P_{max} and N_{max} are fixed at 2000 and 300, respectively.

The constants c_1 and c_2 are used to balance the velocity tendency to the local optimal position P_l and the global optimal position P_g. If c_1 and c_2 are too large, the search space of the particles will be expanded, which may even lead to no convergence of the PSO. If c_1 and c_2 are too small, slow convergence may cause the computation time to be very long. In the case of problems with $P_{max} = 2000$, $N_{max} = 300$, $w = 0.75$, $p_{ms} = 0.6$, $p_{ss} = 0.6$, $p_{cq} = 0.2$, $p_{sq} = 0.2$, 50 trials were separately conducted by varying the values of $c_1 = c_2 \in \{1, 1.5. 2\}$. The average results are summarized in Fig. 3.12.

From Fig. 3.12, when c_1 and c_2 are both set to be 1, the PSO algorithm shows fast convergence and good computational efficiency.

The inertia weight w is set to coordinate the local exploration and the global exploration. If w is too large, there may be a quick local convergence at an early stage of the PSO. If w is too small, the computation time for each iteration will be long, and the optimization rate will become very slow. In the case of problems with $P_{max} = 2000$, $N_{max} = 300$, $c_1 = c_2 = 1$, $p_{ms} = 0.6$, $p_{ss} = 0.6$, $p_{cq} = 0.2$, $p_{sq} = 0.2$, 50 trials were separately conducted by varying the values of $w \in \{0.75, 1, 1.25\}$. The average results are summarized in Fig. 3.13.

From Fig. 3.13, the optimal efficiency and stability are achieved under the condition of $w = 1$.

The problem data (p_{ms}, p_{ss}, p_{cq}, p_{sq}) are determined to help the approach escape from local convergences. Fifty trials are conducted in8 group combinations of the four parameters p_{ms}, p_{ss}, p_{cq}, p_{sq} to illustrate the selection of these parameters. The average *CP*s for the 50 trials are listed in Table 3.28. It is shown that the combination of $p_{ms} = 0.6$, $p_{ss} = 0.6$, $p_{cq} = 0.2$ and $p_{sq} = 0.2$ can yield the best performance.

In conclusion, the performance of the modified PSO for Part 2 is good, and $P_{max} = 2000$, $N_{max} = 300$, $w = 1$, $c_1 = c_2 = 1$, $p_{ms} = 0.6$, $p_{ss} = 0.6$, $p_{cq} = 0.2$, $p_{sq} = 0.2$. The best simulation result and average simulation result are shown in Tables 3.29 and 3.30, respectively.

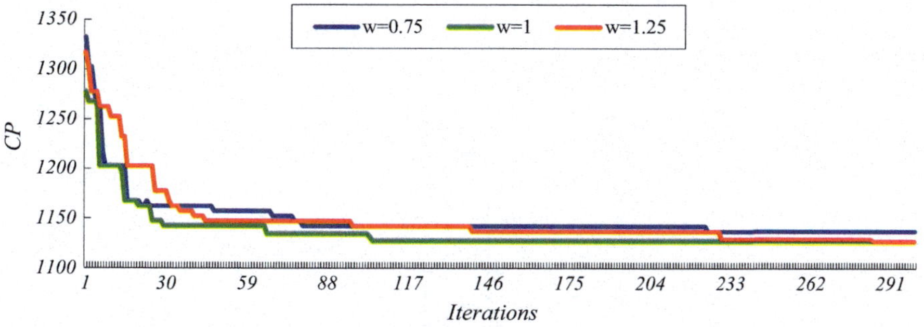

Fig. 3.13 *CP* of the proposed PSO with different inertia weights *w*

The above method of determining the key parameters is based on Part 2. The method of choosing parameters for Part 3 is same as for Part 2. In the case of problems with $P_{max} = 2000$, $N_{max} = 500$, $w = 1.25$, $c_1 = c_2 = 1$, $p_{ms} = 0.6$, $p_{ss} = 0.6$, $p_{cq} = 0.3$, $p_{sq} = 0.3$, 50 trials were separately conducted. The best process plans are listed in Table 3.31, and the average results are listed in Table 3.32.

The following three conditions are used to verify the modified PSO approach for the example part.

(1) All machines and cutting tools are available, and w_1–w_6 in Eq. (3.18) are set as 1.
(2) All machines and cutting tools are available, and $w_1 = w_5 = 0$, $w_1 = w_3 = w_4 = w_6 = 1$.
(3) M_2 and T_7 are down, $w_2 = w_5 = 0$, $w_1 = w_3 = w_4 = w_6 = 1$.

For Part 2, 50 trials were performed under conditions (a) and (b). A penalty cost is included in the *CP* to facilitate the comparison with other approaches, which is 200 for Part 2 [4, 7]. A comparison with the results obtained using the GA and SA approaches [2], TS [4], and HBMO [14], as well as the two-stage ACO [7], is provided in Table 3.33.

Under condition (a), this approach is same as SA, TS, HBMO, and two-stage ACO and is better than GA using the minimum machine costs. Using the maximum machine costs, the *CP* (1328.0) is the same as that of HBMO, and it is better than the other four approaches. This approach has the best performance in the mean machine cost (1328). It is obvious that the same CP (1328.0) is achieved 50 times and is superior to all of the other approaches.

Under condition (b), the performance of this approach is the same as those of HBMO and the two-stage ACO. Using the minimum machine costs, this approach is better than GA, and it is the same as the other four approaches. Using the maximum and mean machine costs, this approach is better than GA, SA, and TS.

For Part 3, 50 trials were carried out under conditions (a), (b) and (c). A comparison of the results with those of TS [4], PSO [9], and HBMO [14], as well as the two-stage ACO [7] is provided in Table 3.34.

Table 3.28 Determination of four probabilities of the modified PSO

	(Mutation1: p_{ms}, Mutation2: p_{ss}, Crossover: p_{cq}, Shift: p_{sq})							
	(0.6, 0.4, 0.2, 0.4)	(0.4, 0.6, 0.4, 0.2)	(0.6, 0.6, 0.4, 0.4)	(0.6, 0.6, 0.2, 0.2)	(0.4, 0.4, 0.2, 0.2)	(0.4, 0.4, 0.4, 0.4)	(0.4, 0.6, 0.2, 0.4)	(0.6, 0.4, 0.4, 0.2)
Mean	1198.5	1198.8	1308.4	1131.8	1136.6	1541.3	1206.1	1186.7
Maximum	1263	1318	1713	1163	1188	1998	1318	1263
Minimum	1143	1143	1148	1128	1128	1158	1143	1143

Table 3.29 Best simulation result for Part 2

Operation	Machine	Tool	TAD	Result
6	2	2	− Z	$CM = 490$ $CT = 98$ $CTC = 60$ $CS = 480$ $CP = 1128$
1	2	1	− Z	
7	2	1	− Z	
9	2	1	− Z	
12	2	1	− Z	
5	2	5	− Z	
3	2	5	+ Y	
4	2	5	+ Y	
8	2	5	+ X	
10	2	5	− Y	
11	2	5	− Y	
13	2	5	− Y	
14	2	1	− Y	
2	2	8	− Y	

Table 3.30 Average simulation result of 50 trials for Part 2

Type	Mean	Maximum	Minimum	Standard deviation
CM	490	490	490	0
CT	98.2	103	98	0.98
CMC	0	0	0	0
CTC	60.9	75	60	3.56
CS	480	480	480	0
CP	1128.9	1143.0	1128	3.56

Under condition (a), the minimum machine cost is the same as that of HBMO and the two-stage ACO, and it is better than TS and PSO. Among the 50 trial results, *CP* (2525) occurs 23 times, *CP* (2527) occurs 20 times, and *CP* (2535) occurs 7 times. Accordingly, this approach is superior to all of the other approaches on the mean machine cost. Under condition (b), the minimum machine cost (2090) is the same as that of HBMO and the two-stage ACO. *CP* (2090) occurs 45 times, and *CP* (2120) occurs 5 times in 50 trials; the mean machine cost is the best among all of the approaches. Under condition (c), *CP* (2590) occurs 34 times, and *CP* (2600) occurs 16 times in 50 trials. Generally, the performance of this approach is similar to that of HBMO under condition (c) and is superior to those of the two-stage ACO and TS.

Table 3.31 Best simulation result for Part 3

Operation	Machine	Tool	TAD	Result
1	2	6	+Z	$CM = 770$ $CT = 235$ $CMC = 320$ $CTC = 200$ $CS = 1000$ $CP = 2525$
3	2	6	+X	
5	2	6	+X	
6	2	6	−Z	
2	2	6	−Z	
18	2	6	−Z	
11	2	7	−Z	
12	2	2	−Z	
13	2	9	−Z	
17	2	7	−X	
7	2	7	−a	
8	2	2	−a	
9	2	9	−a	
19	2	9	+Z	
14	4	10	−Z	
20	4	10	+Z	
10	4	10	−a	
4	1	2	−Z	
15	1	1	−Z	
16	1	5	−Z	

Table 3.32 Average simulation result of 50 trials for Part 3

Type	Mean	Maximum	Minimum	Standard deviation
CM	770	770	770	0
CT	240	267	235	10.13
CMC	320	320	320	0
CTC	197.2	180	200	6.94
CS	1000	1000	1000	0
CP	2527.2	2535.0	2525.0	3.28

3.6 Conclusions

Firstly, a two-stage ACO approach is proposed to solve the PP problem for prismatic parts. A directed graph is used to represent PP problem. The graph consist of nodes, directed/undirected arcs, OR relations. Artificial ants travel on the graph freely to construct process plans in two stages. The operation nodes are selected in the first stage, which will reduce the graph to a simple graph. In the second stage, the selected

Table 3.33 Results compared to other approaches

Condition	Proposed approach	GA	SA	TS	HBMO	Two-stage ACO
(a)						
Mean	1328	1611.0	1373.5	1342.6	1328	1329
Maximum	1328	1778	1518	1378	1328	1348
Minimum	1328	1478	1328	1328	1328	1328
(b)						
Mean	1170	1482	1217	1194	1170	1170
Maximum	1170	1650	1345	1290	1170	1170
Minimum	1170	1410	1170	1170	1170	1170

Table 3.34 Results compared to other approaches

Condition	Proposed approach	TS	PSO	HBMO	Two-stage ACO
(a)					
Mean	2527.2	2609.6	2680.5	2543.5	2552.5
Maximum	2535	2690	–	2557	2557
Minimum	2525	2527	2535	2525	2525
(b)					
Mean	2093.0	2208.0	–	2098.0	2120.5
Maximum	2120	2390	–	2120	2380
Minimum	2090	2120	–	2090	2090
(c)					
Mean	2593.2	2630.0	–	2592.5	2600.8
Maximum	2600	2740	–	2600	2740
Minimum	2590	2580	–	2590	2590

operations on the simple graph will be sequenced to generate the feasible process plans. As evident from the experimental results, the two-stage approach is able to provide a feasible solution method for the PP problem.

Secondly, a traditional PSO approach is modified to solve the PP problem. Efficient encoding and decoding, population initialization, and iteration and control methods have been designed. Meanwhile, to avoid local convergence, several new operators for the different parts of the particles have been designed and incorporated into the traditional PSO to improve the particles' movements. Simulation experiments show that the modified PSO algorithm can perform the process plan optimization competently and consistently and generate better solutions compared with other approaches.

In the simulation experiment, a small change in the parameters induces computational result saltation. Hence, a deep discussion of selecting the modified PSO approach parameters will be conducted. Additionally, integrated process planning and scheduling that considers the better performance of manufacturing systems may be a direction for further study.

Acknowledgements This work was supported by the Fundamental Research Funds for the Central Universities (14DZ37). The authors would like to thank all the people who may help to improve the manuscript.

References

1. Zhang F, Zhang YF, Nee AYC (1997) Using genetic algorithms in process planning for job shop machining. IEEE Trans Evol Comput 1(4):278–289
2. Li WD, Ong SK, Nee AYC (2002) Hybrid genetic algorithm and simulated annealing approach for the optimization of process plans for prismatic parts. Int J Prod Res 40(8):1899–1922
3. Ma GH, Zhang YF, Nee AYC (2000) A simulated annealing-based optimization algorithm for process planning. Int J Prod Res 38(12):2671–2687
4. Li WD, Ong SK, Nee AYC (2004) Optimization of process plans using a constraint-based tabu search approach. Int J Prod Res 42(10):1955–1985
5. Lian KL, Zhang CY, Shao XY, Zeng YH (2011) A multi-dimensional tabu search algorithm for the optimization of process planning. Sci China-Technol Sci 54(12):3211–3219
6. Liu XJ, Yi H, Ni ZH (2013) Application of ant colony optimization algorithm in process planning optimization. J Intell Manuf 24(1):1–13
7. Wang JF, Wu XH, Fan XL (2015) A two-stage ant colony optimization approach based on a directed graph for process planning. Int J Adv Manuf Technol 80(5–8):839–850
8. Hu QW, Qiao LH, Peng GW (2015) An ant colony approach to operation sequencing optimization in process planning. Proc Inst Mech Eng B: J Eng manufact. https://doi.org/10.1177/0954405415616786
9. Guo YW, Mileham AR, Owen GW, Li WD (2006) Operation sequencing optimization using a particle swarm optimization approach. Proc Inst Mech Eng B: J Eng Manufact 220(12):1945–1958
10. Kafashi S, Shakeri M, Abedini V (2012) Automated setup planning in CAPP: a modified particle swarm optimisation-based approach. Int J Prod Res 50(15):4127–4140
11. Wang YF, Zhang YF, Fuh JYH (2012) A hybrid particle swarm based method for process planning optimisation. Int J Prod Res 50(1):277–292
12. Li XY, Gao L, Wen XY (2013) Application of an efficient modified particle swarm optimization algorithm for process planning. Int J Adv Manuf Technol 67(5–8):1355–1369
13. Miljković Z, Petrović M (2016) Application of modified multi-objective particle swarm optimisation algorithm for flexible process planning problem. Int J Comput Integr Manuf. https://doi.org/10.1080/0951192X.2016.1145804
14. Wen XY, Li XY, Gao L, Sang HY (2014) Honey bees mating optimization algorithm for process planning problem. J Intell Manuf 25(3):459–472
15. Huang WJ, Hu YJ, Cai LG (2012) An effective hybrid graph and genetic algorithm approach to process planning optimization for prismatic parts. Int J Adv Manuf Technol 62(9–12):1219–1232
16. Kennedy J, Eberhart RC (1995) Particle swarm optimization. In: Proceedings of the 1995 IEEE international conference on neural networks, vol 4. IEEE Press, Piscataway, NJ, pp 1942–1948
17. Hrelja M, Klancnik S, Irgolic T, Paulic M, Jurkovic Z, Balic J, Brezocnik M (2014) Particle swarm optimization approach for modelling a turning process. Adv Prod Eng Manage 9(1):21–30
18. Wang Y, Liu JH (2010) Chaotic particle swarm optimization for assembly sequence planning. Robot Comput Integr Manufact 26(2):212–222

19. Malik J, Mishra R, Singh I (2011) PSO-ANN approach for estimating drilling induced damage in CFRP laminates. Adv Prod Eng Manage 6(2):95–104
20. Dorigo M, Maniezzo V, Colorni A (1996) The ant system: Optimization by a colony of cooperating agents. IEEE Trans Syst Man Cybern B Cybern 26(1):29–41
21. Montemanni R, Gambardella LM, Rizzoli AE, Donati AV (2005) Ant colony system for a dynamic vehicle routing problem. J Comb Optim 10:327–343
22. Hamzheei M, Farahani RZ, Rashidi-Bajgan H (2013) An ant colony-based algorithm for finding the shortest bidirectional path for automated guided vehicles in a block layout. Int J Adv Manuf Technol 64:399–409
23. Ghaiebi H, Solimanpur M (2007) An ant algorithm for optimization of hole-making operations. Comput Ind Eng 52:308–319
24. Liu XJ, H Y, Ni ZH, Qi JC, Qiu XL, (2013) Process planning optimization of hole-making operations using ant colony algorithm. Int J Adv Manuf Technol 69:753–769
25. Zheng QX, Li M, Li YX, Tang QH (2013) Station ant colony optimization for the type 2 assembly line balancing problem. Int J Adv Manuf Technol 66:1859–1870
26. Yu JP, Wang CG (2013) A max-min ant colony system for assembly sequence planning. Int J Adv Manuf Technol 67:2819–2835
27. Prakash A, Tiwari MK, Shankar R (2008) Optimal job sequence determination and operation machine allocation in flexible manufacturing systems: An approach using adaptive hierarchical ant colony algorithm. J Intell Manuf 19:161–173
28. Korytkowski P, Rymaszewski S, Wisniewski T (2013) Ant colony optimization for job shop scheduling using multi-attribute dispatching rules. Int J Adv Manuf Technol 67:231–241
29. Krishna AG, Mallikarjuna Rao K (2006) Optimisation of operations sequence in CAPP using an ant colony algorithm. Int J Adv Manuf Technol 29:159–164
30. Liu XJ, Yi H, Ni ZH (2013) Application of ant colony optimization algorithm in process planning optimization. J Intell Manuf 24:1–13

Chapter 4
Intelligent Workshop and Scheduling

4.1 Intelligent Workshop

Intelligent manufacturing integrates the advanced technologies such as modern sensing technology, network technology, and automation technology. A large number of workshop production data has been collected through intelligent perception, human–computer interaction. These data involve product demand design, raw materials procurement, manufacturing, warehousing, sales and other links. According to the standards of GB/T20720.3-2010, digital workshops mainly include digital manufacturing and auxiliary equipment for production operation management, quality operation management, maintenance operation management, etc. The intelligent workshop is a "smart" workshop with automatic sensing and artificial intelligence based on manufacturing IoT and manufacturing data cloud. According to the system framework of intelligent manufacturing system, the architecture of intelligent workshop can be summarized as equipment perception layer—data transmission layer—analysis processing layer—application service layer, which is shown in Fig. 4.1.

1. **Equipment perception layer**

Real-time multi-source data such as status, operation, process and other real-time multi-source data generated by various production elements such as equipment, workers, materials, tools, and work-in-progress in the workshop are the basis of production process optimization and control. According to the characteristics of different production factors and the requirements of data collection and application, various sensing devices such as RFID, sensors, and ultra-wideband (UWB) are configured on the workshop to realize interconnection, mutual induction and data collection of various production factors, which can ensure real-time and reliable acquisition of multi-source information in the workshop. The communication technologies used for short-distance data transmission include Bluetooth technology, infrared technology, WiFi technology, and the advanced wireless sensor network

J. Wang, *Intelligent Manufacturing System and Intelligent Workshop*,
Advanced and Intelligent Manufacturing in China,
https://doi.org/10.1007/978-981-99-2011-2_4

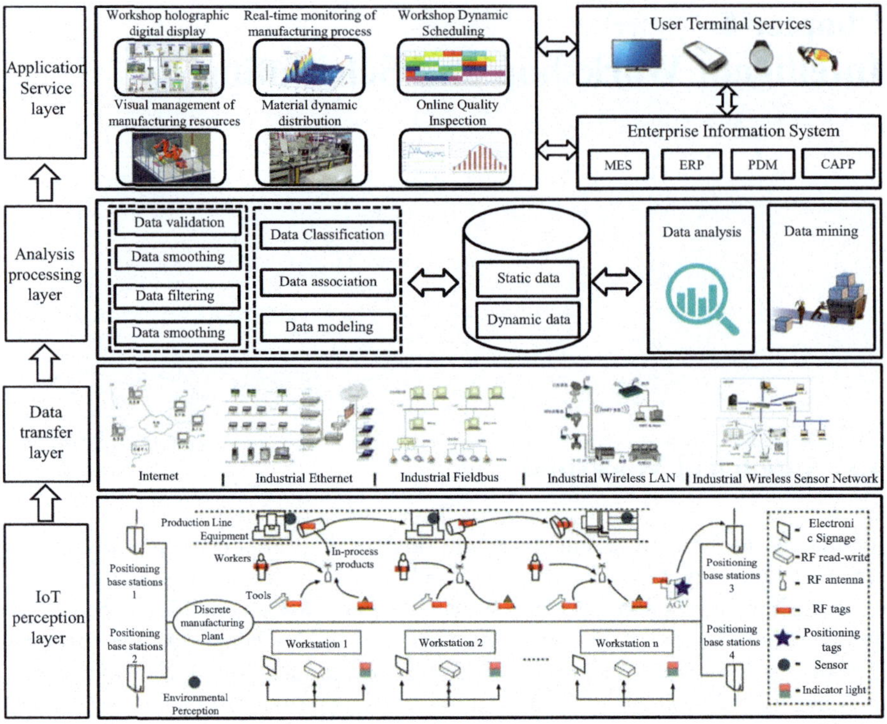

Fig. 4.1 Basic architecture of intelligent manufacturing workshop

based on the ZiBee protocol. This layer is responsible for the collection of multi-source information in the manufacturing process. Therefore, two problems need to be solved, namely, the optimal configuration of the sensor network and the modelling of manufacturing resources. The manufacturing information involved in the manufacturing process is complex and multi-resource, such as the movement information and status information of equipment, workers, work in progress, materials, tools, etc. At the same time, the traditional wired network solution and the network solution based on wireless AP (access point) are not suitable for the transmission of dynamic manufacturing information in a complex workshop due to the limitations of the workshop space, strong mobility of manufacturing resources, and communication blind spots. Therefore, it is necessary to choose a heterogeneous multi-hop network (WSN) with dynamic self-organizing network and maximum possible elimination of blind spots to realize reliable transmission of dynamic manufacturing information. In order to realize the automatic perception and cross-connection of manufacturing resources, it is necessary to set the various sensing devices on the corresponding manufacturing resources. These sensing devices make the manufacturing resources have certain logical behavior capabilities, can actively perceive changes in the manufacturing environment, and can also perceives the real-time operating status and environmental change data of the manufacturing resources. This new manufacturing object, which is composed of traditional manufacturing resources and advanced sensing equipment, can be defined as intelligent manufacturing resources. According to the different functions of the sensing devices, intelligent manufacturing resources are divided into intelligent manufacturing resources with perception capabilities and intelligent manufacturing resources that are perceived. Taking RFID sensing equipment as an example, workshop staff equipped with RFID tags, and manufacturing equipment equipped with RFID readers and sensors can all be regarded as intelligent manufacturing resources. Among them, manufacturing resources equipped with RFID readers can be regarded as intelligent manufacturing resources with perception capabilities which can sense the perceivable intelligence equipped with RFID tags within a certain distance.

With the modeling idea of Agent, intelligent manufacturing resources can implement their own business logic and interact and cooperate with other intelligent manufacturing resources according to the predefined workflow model, and perceive and analyze possible or certain dynamic conditions in the manufacturing environment. When different sensing devices are placed on different manufacturing resources according to the measurement acquisition requirements, the management of the data acquisition events of these sensing devices becomes very important. Since the Agent has the characteristics of autonomy, initiative and intelligence, each sensor device is managed based on Agent technology. The Agent can implement its own business logic according to the predefined work logic, so as to sense and analyze the environment changes in the manufacturing workshop environment, and actively obtain the change information of the manufacturing environment. The management architecture of sensing devices based on Agent technology is shown in Fig. 4.2, where each sensing device corresponds to an Agent related to its behaviour. The Agent adopts the architecture of Web Service, and establishes the relationship between the Agent and the sensing device. The information reading behaviour of the device is controlled, and then the real-time production information collected is transmitted to external users in the form of Web Service.

Agent's service function is described as follows:

(1) Workflow model. The real-time manufacturing information from the multiple sensing devices placed on the manufacturing equipment is different, such as time domain and information collection sequence. In order to flexibly configure the collaborative workflow of different sensing devices, workflow technology is usually used to establish the working order and trigger conditions of each sensing device, which can realize the collaborative work on complex real-time manufacturing information collection.
(2) Bind the model. It is used to bind the affiliation relationship between sensing equipment and manufacturing resources, which build the foundation for the seamless connection of collected real-time production information and production scheduling system, such as the relationship between sensing equipment and manufacturing equipment, and the manufacturing tasks assigned to the manufacturing equipment.

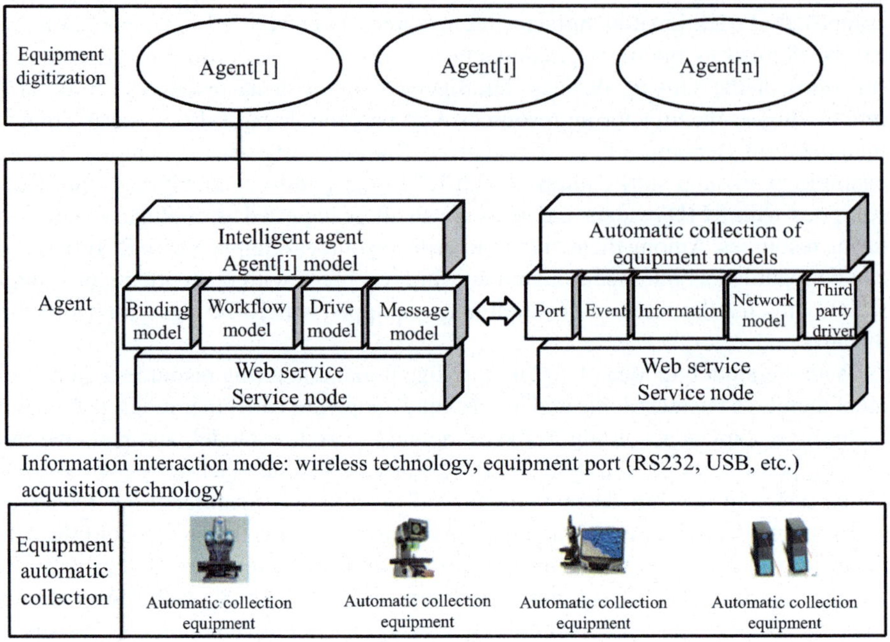

Fig. 4.2 Agent-based smart device management model

(3) Drive model. At present, two modes are adopted: to control the sensing device by standard interfaces such as serial port, USB port, Bluetooth, etc.; to realize the control of the device directly by packaging the driver of the sensing device such as dynamic link library, etc.

(4) Message model. The extensible markup language (eXtensible Markup Language, XML) standard template is used to encapsulate the real-time information collected by the sensing device, and the real-time information is transmitted through the Web Service through the Simple Object Access Protocol (SOAP).

2. **Data transfer layer**

In the workshop, data such as production status, material flow, environmental parameters, equipment operation, and quality inspection are distributed widely. According to the different data transmission characteristics and requirements of different sensing devices, these data are selectively transmitted through the Internet, industrial Ethernet, field bus, industrial local area network, industrial sensor network, etc. The transmission approaches realize the effective transmission and exchange of sensory information, and ensure the stable transmission and application of production data in the workshop. On the one hand, the running data is transmitted to the data analysis and processing layer through the SCADA system. On the other hand, the control data including various control commands issued by DCS, PLC and FCS systems are transmitted to various intelligent devices in the workshop. The intelligent equipment in the intelligent workshop communicates by computer technology, communication technology, and numerical control technology, which complete the network coverage of the digital workshop and realize the centralized management and control of the intelligent equipment in the workshop. Early communication methods were mainly point-to-point communication based on asynchronous serial communication. With the development of computer technology, network technology and field bus technology, communication methods based on field bus and Ethernet gradually developed.

(1) Communication method based on serial port. The point-to-point communication method is the earliest communication method used in industrial equipment networking. It is realized based on RS232C/RS422 serial port. Its topology is star-shaped, and the communication rate is generally 110–9600 bit/s. This point-to-point connection method is simple and low in cost. Since most computers and smart devices have serial communication interfaces, it is relatively convenient to implement. However, this connection has disadvantages such as short transmission distance, unreliable transmission, low transmission rate, poor real-time performance, and difficult system expansion. In order to overcome the above shortcomings, people have proposed various technical means to meet the development needs of communication networking. The first is to realize the communication with multiple devices through the serial port expansion card. The second is to directly connect to the network through a serial port server. The serial port server in the equipment network is to realize the conversion between the serial port data flow in the bottom layer RS232 link and the upper layer Ethernet port data packet based on the TCP/IP protocol.
(2) Communication mode based on LAN. With the development of network technology and automation technology, the network interface has gradually become the standard configuration of modern intelligent equipment. In addition, the intelligent equipment with only serial interfaces such as RS232 or RS485 can also be converted into a network interface through a serial port server. All of these have laid the foundation for the communication connection based on the Local area network (LAN). LAN communication is a kind of non-centralized control communication network, which connects scattered intelligent equipment through a common communication medium such as twisted pair, optical fiber cable or coaxial cable. LAN communication has the characteristics of high reliability, high speed transmission of information and data sharing in a local area. Common LANs include MAP network, industrial Ethernet, etc. Due to the high complexity and high development cost of MAP network in actual development, industrial Ethernet is the most popular at present. Industrial Ethernet is a LAN technology that integrates advanced distributed control technology on the basis of traditional Ethernet technology, so that it can be applied to industrial control and management. The network layer and transport layer of industrial Ethernet are mainly based on TCP/IP. The protocol is characterized by open protocol standards, independent of specific computer hardware and operating

systems, and has a unified network address allocation scheme and standardized high-level protocols. In the digital workshop site, industrial Ethernet technology is used to build a communication network to meet the real-time communication requirements of all levels of the workshop and the needs of information integration with the upper-level enterprise.

(3) Fieldbus-based communication mode. Under the consideration of protecting the own investment interests, fieldbus is used to implement the network connection of field-level intelligent equipment in the application of the control level and above levels. Fieldbus is a control network with multiple network segments, multiple communication media and multiple communication rates. It is mainly used in production sites to realize two-way, serial, and multi-node communication between field devices, between field devices and control devices. It can not only transmit the operating parameters, status information and fault information of a field device, but also send various control and configuration commands to related field devices. The fieldbus topology mainly adopts the bus type. In the communication mode, the data exchange between the nodes is only allowed to be realized through the master node, and the data is not directly exchanged between the nodes. This helps simplify communication, reduce costs, and improve reliability. In the automatic control system, the fieldbus such as CANBUS has been widely used in the interconnection of the bottom equipment in the workshop due to its low cost, strong anti-interference ability, and good real-time performance.

The networking communication technologies of the intelligent workshop involve serial communication, local area network communication and fieldbus. In the on-site networking communication structure, the fieldbus network is mainly used in the underlying communication network. In the actual information transmission process of the industrial control underlying network, the amount of control-oriented information for a single node is not large, and the transmission task is relatively simple. But the requirements of the rapid information transmission are relatively high. As a real-time control network, fieldbus can realize real-time, stable, reliable, complete and accurate communication between production field equipment because of its high stability and strong anti-interference ability characteristics. Therefore, the CANBUS bus is considered as the communication networking method of the underlying intelligent equipment, and each intelligent equipment is connected to the communication network as a CAN node. In summary, the on-site networking communication structure of the intelligent workshop is shown in Fig. 4.3.

The analysis and processing layer is the connecting position in the manufacturing execution system (MES). It is the hub for data connection and communication between the upper application service layer and the underlying intelligent equipment. The two types of network interaction modes are adopted in the network communication between the MES business management layer and the analysis and processing layer, that is, the C/S mode based on the client and the server and the B/S mode based on the browser and the WEB server. The C/S network mode is the most commonly used communication mode at present. In the C/S network interaction mode, the client application program is designed for specific tasks, and has the characteristics

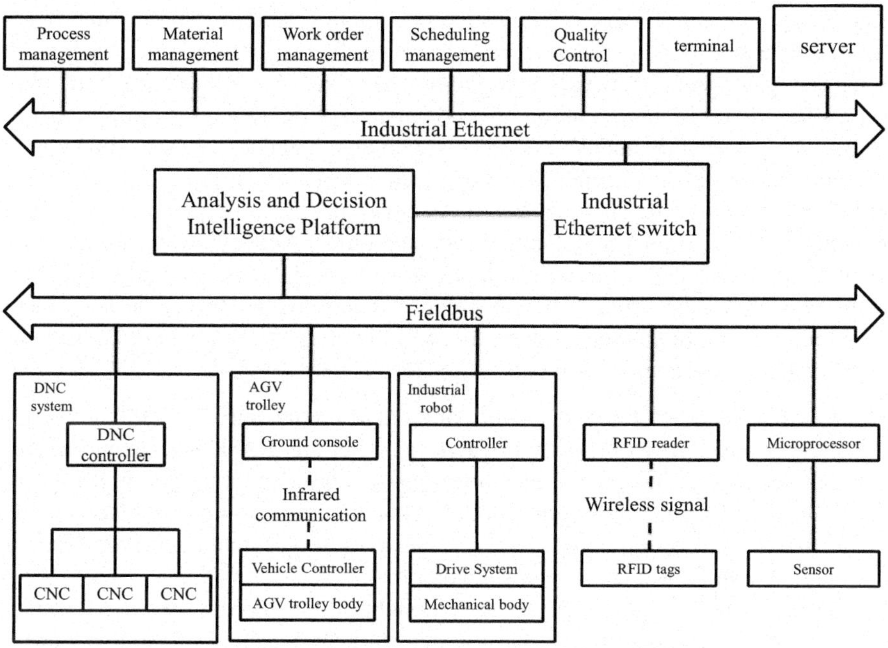

Fig. 4.3 On-site networking communication structure of intelligent workshop

of outstanding specialization and software personalization. At the same time, the network traffic of the C/S structure only includes the communication between the client and the server. The network communication volume is low, and the response speed is fast, which is conducive to processing a large amount of data. In the B/S network mode, the system puts most of the business logic on the server side, which requires high performance of the server, and the "crash" of the server will cause all clients to fail to work normally. At the same time, the system adopts an open structure mode. The security can only rely on the user account management module on the server, which usually results in the poor data security. So, the C/S network mode is more in line with the requirements of the upper layer network communication of the intelligent equipment integration platform.

3. **Analysis processing layer**

In the manufacturing workshop there are complex environmental characteristics such as strong metal interference, occlusion and coverage, as well as complex production characteristics such as multi-variety variable batch mixed-line production, variable production conditions, and movement of production factors, which lead to redundancy, disorder of manufacturing data and strong uncertainty. It is necessary to convert the manufacturing data with typical characteristics of large capacity and low value density into the effective data used in production and management by smoothing, filtering, fusion, classification, association and other processing operations. The analysis and processing layer based on the big data information of workshop manufacturing is currently recognized as a typical feature of intelligent workshops. Based on big data thinking, the information of equipment status parameters, plan execution, quality and delivery date are digitized, and the relationship between the manufacturing data through clustering, sequential pattern mining, and association algorithms is analyzed; Then, the evolution law of the performance such as delivery punctuality rate, product qualification rate, equipment status and operation process is obtained using data mining; finally, based on control theory, find key parameters from the evolution law for quantification control to ensure that the performance meets the requirements. In the above ideas, a new decision-making model of "correlation + prediction + regulation" driven by big data has been formed.

(1) Correlation refers to the discovery of hidden relationships through the correlation analysis of workshop manufacturing data. On the basis of manufacturing data preprocessing such as cleaning, classification, and integration, it is necessary to construct a time series model of manufacturing data and mine sequence patterns, realize the correlation analysis of different manufacturing data, and mine the influence rules between data.
(2) Prediction refers to describing the internal relationship between the workshop manufacturing process and performance indicators using of the above correlation analysis results. A model to describe the influence law of the workshop running process data on the performance index data is established to realize the prediction of workshop performance.
(3) Regulation refers to finding and controlling the key manufacturing parameters of the workshop running process based on the prediction model. By determining the key parameters that affect quality control and delivery time control, a scientific control mechanism for workshop performance such as product qualification rate and delivery on-time rate is established using the knowledge of laws.

According to the above decision-making model mentioned, solve intelligent workshop running analysis and decision-making using big data requires the realization of data preprocessing and analysis, running analysis and performance prediction. However, the structured data only can be analyzed on the most of the current workshop data processing and analysis. With the increasing proportion of semi-structured and unstructured data in intelligent workshop data under the big data environment, it is difficult to realize the effective classification and reuse of massive high-dimensional data. The multi-scale characteristics of data time-varying law, the complexity and diversity of the data correlation promote the further in-depth research of the preprocessing and analysis methods of manufacturing data. Therefore, the intelligent workshop analysis and decision-making method of "association + prediction + regulation" realizes the "association" through the manufacturing data preprocessing method, the manufacturing data time series analysis method and the manufacturing data relationship network modeling method, and realizes the "prediction" through the operation state prediction method, to achieve "regulation" through the operation decision-making method.

(1) Manufacturing big data preprocessing method: The manufacturing data generated during the running of the workshop has the characteristics of massive, high-dimensional, multi-source heterogeneous, multi-scale, and high noise. These data are difficult to be directly used for analysis and decision-making. The big data preprocessing method mainly aims to provide reliable and reusable data resources for the analysis and decision-making of workshop running by performing operations such as cleaning and denoising, modeling integration, and multi-scale classification.
(2) Manufacturing big data time series analysis method: Manufacturing big data time series analysis method aims to establish a time series model of multi-dimensional data in the workshop running process, designs a time series pattern mining algorithm of manufacturing data, and reveals changes law of manufacturing data.
(3) Manufacturing big data relational network modeling method: Manufacturing data such as products, processes, equipment, and system running interact with each other, which makes the workshop production process present complex running characteristics. The workshop manufacturing big data relational network modeling method is based on the association analysis of manufacturing data such as process parameters and equipment status parameters.
(4) Workshop running state prediction method: The workshop running state prediction method aims to analyze the dynamic characteristics and running mechanism of the internal structure of the workshop manufacturing system according to the time series model of manufacturing data.
(5) Decision-making method for workshop running status: Based on the analysis of workshop running, workshop performance is optimized through scientific adjustment of key manufacturing data.

4. **Application service layer**

Perception data is used for manufacturing workshop management and production process control optimization. The functional services such as holographic digital display of workshops, visual management of manufacturing resources, real-time monitoring of manufacturing processes, dynamic material distribution, dynamic production scheduling and quality diagnosis and traceability are also provided in this layer. The close integration with information systems such as MES, ERP, and Product Data Management (PDM) is realized through a unified data integration interface.

4.2 An Application of Genetic Algorithms for the Flexible Job-Shop Scheduling Problem

4.2.1 Introduction

Most scheduling problems are complex combinatorial optimization problems and very difficult to solve. The process constraints and shop environment are important to find the best scheduling in the planning and managing of manufacturing processes. The Job-Shop Scheduling Problem (JSP) is one of the most classical issues in this area, which may be described as follows: each one of n jobs must be processed passing through m machines in a given sequence. The sequence of m machines is different for each different job. When one job was processed on a machine, it can be considered as one operation, each job needs a combination of m operations to complete the work. There are several constraints on jobs and machines. (1) Only one operation can be processed on one of m machines at a time, and cannot be changed during the processing. (2) Each machine can only process one operation at the same time. (3) Any job can not interrupt the machine which is processing one operation of another job. The main objective of the JSP is to find an optimal operations schedule to minimize the maximum completion time, i.e., makespan ($C_{\max}$). The JSP is a well-known NP-hard problem.

Although a successful algorithm for the classical JSP has not been developed, there is a trend in the research domain to solve a much more complex version of the problem, i.e., the Flexible Job-shop Scheduling Problem (FJSP) which allows an operation to be processed on one out of several machines. It is more complex and difficult than the JSP because the assignment of operations to machines needs to be determined additionally. The scheduling problem of a FJSP consists of a two sub-problem. (1) Routing sub-problems. It assigns each operation of all jobs to a machine out of a set of capable machines. (2) Scheduling sub-problem. It sequences the assigned operations to be processed on all machines in order to obtain a viable and effective schedule minimizing an objective function. Therefore, The FJSP mainly needs to finish two works. The first one is to assign each operation to a machine, and the second one is to schedule these operations in order to make an objective function minimal [1], which are accomplished independently from each other.

Bruker and Schlic address the FJSP in 1990. They proposed a polynomial algorithm for solving the FJSP with two jobs. Integrated approaches and hierarchical approaches have been used. In integrated approaches, assignment of operations to machines and the sequencing of operations on the resources or machines are treated in the same way. In order to reduce the FJSP complexity, hierarchical approaches decomposed the original problem, namely, assignment and sequencing are considered independently and treated separately. This type of approach is natural for FJSP since the routing and the scheduling sub-problems can be separated [2]. Brandimarte was the first to apply the hierarchical approach into the FJSP [3].

Paulli used some existing dispatching rules to solve the routing sub-problem, and then solved the scheduling sub-problem by different Tabu search methods.

Although integrated approach is difficult to be implemented in engineering practice, it could achieve better results [4]. Wang et al. developed a filtered-beam-search-based heuristic algorithm to find sub-optimal schedules for the FJSP, The proposed algorithm incorporated dispatching rules based heuristics and explores the search space, which could avoid useless paths and improve the search speed [5].

Kacem et al. proposed a genetic algorithm which uses tasks sequencing list coding scheme. In this coding scheme a chromosome is represented by combining both routing and sequencing information, and an approach is developed by localization (AL) to find promising initial assignment [6, 7]. Lei developed an efficient decomposition-integration genetic algorithm to minimise the maximum fuzzy completion time. The proposed algorithm decomposed the chromosomes of the main population into a job sequencing part and a machine assigning part and independently evolved the populations [8].

The heuristic of considering assignment and scheduling at the same time were used widely to solve the FJSP. Hurink et al. proposed a taboo search heuristic in which reassignment and rescheduling are considered as two different types of moves. Based on the integrated approach which had been presented by Dauzere-Peres and Paulli, a neighbourhood structure for the problem where there is no distinction between reassigning and resequencing an operation was defined, and a taboo search procedure is proposed based on the neighbourhood structure [9, 10]. Mastrolilli and Gambardella also improved Dauzere-Peres' taboo search techniques and presented two neighbourhood functions [11]. Amiri et al. propose a variable neighbourhood search algorithm that solves the FJSP to minimise makespan [12].

Lu proposed a genetic algorithm to solve a distributed and flexible job-shop scheduling problem. A new and concise chromosome representation models a 3-dimensional scheduling solution by a 1-dimensional scheme [13]. Many researchers were interested in applying hybrid genetic algorithms to solve the FJSP [14, 15]. Kalinowski presents an application of ant colony optimisation algorithm for scheduling multi-resources operations in flexible job shop type of production systems [16]. Nouiri study the distribution of the PSO-solving method for future implementation on embedded systems that can make decisions in real time according to the state of resources and any unplanned or unforeseen events [17]. Recently, bee colony algorithm, gray wolf approach and the other heuristic were also used to solve the FJSP, as proven by the growing number of papers on the topic [18–20].

The flexible job-shop scheduling problem can be formulated as follows.

The flexible job shop scheduling (FJSP) problem can be described as follows: m machine tools $\{W_1, W_2, \ldots W_m\}$ complete n jobs $\{J_1, J_2, \ldots J_n\}$, where each job J_i consists of p processes $\{O_{i1}, O_{i2}, \ldots O_{ip}\}$, the number and sequence of each operation are certain. Each process can be completed by multiple machine tools, and the processing time varies with different machines. The goal of scheduling is to arrange each process ($n \times p$) of n tasks to be processed on the corresponding machine tools, determine the optimal sequence and start time of the corresponding operations of each machine tool, and minimize the completion time of all tasks. Based on the needs of the problem description and the actual situation of the project, FJSP also needs to introduce the following assumptions:

(1) The machine tool W_k that completes the operation O_{ij} is optional, and the processing time T_{ijk} of the specified machine tool for a certain operation is known, where $\{O_{ij} \in O, W_k \in W, i = 1, 2, \ldots, n, j = 1, 2, \ldots, p, k = 1, 2, \ldots, m\}$;
(2) At a certain moment, each machine tool W_k can only handle one job J_i, the two operations of the same job cannot be processed at the same time, and the operation O_{ij} cannot be interrupted once started;
(3) The machine adjustment setting time and the transportation time of the operation are ignored;
(4) Regardless of job cancellation, machine crashes and other random factors.

Therefore, FJSP can be expressed as:

$$S = \{O_{111}, \ldots, O_{11k}, O_{121}, \ldots O_{12k}, \ldots, O_{1jk}, \ldots, O_{2jk}, \ldots O_{ijk}\} \tag{4.1}$$

In the Eq. (4.1), O_{ijk} means that the jth operation of the ith job is processed on the kth machine tool.

An example of the 6X6 FJSP is listed in Table 4.1. This example consists of 6 jobs, each job consists of 3 operations. 18 operations of the 6 jobs are assigned to 6 machine tools for processing. The data in the table indicates the processing time of each operation when it is processed on the corresponding machine tool, and "X" indicates that the operation cannot be processed on the corresponding machine tool. For example, process O_{11} can be completed by five machine tools W_1, W_2, W_4, W_5 and W_6, and the processing time is 2 s, 3 s, 5 s, 2 s and 3 s respectively.

The main research purpose on FJSP is operations sequencing and machine tool selecting. For different processing requirements, the final optimization objectives are slightly different. Usually, the performance indicators of FJSP problems include the following categories: indicators based on the completion time of all jobs, indicators based on delivery time, indicators based on cost, and indicators based on machine tool load.

1. **Indicator based on the completion time of processing jobs**

The completion time of the last operation of each job is called the completion time of the job, and the latest completion time of all jobs is the maximum completion time (makespan). Minimizing the maximum completion time is the ultimate goal of improving the processing efficiency of the jobs, which is the most important indicator of the FJSP problem. Usually, Eq. (4.2) is used as the evaluation index based on completion time.

$$f_1 = \min(C_{\max}) \tag{4.2}$$

In Eq. (4.2), $C_{\max} = \max(C_j)$, C_J is the completion time of job J_i.

2. **Indicators based on delivery time**

At present, the multi-variety and variable batch production mode has gradually become the main form of enterprise organization production. In this mode, how to

Table 4.1 An example of 6X6 FJSP

Job	Operation	Machine					
		W_1	W_2	W_3	W_4	W_5	W_6
1	O_{11}	2	3	X	5	2	3
	O_{12}	4	3	5	X	3	X
	O_{13}	2	X	5	4	X	4
2	O_{21}	3	X	5	3	2	X
	O_{22}	4	X	3	3	X	5
	O_{23}	X	X	4	5	7	9
3	O_{31}	5	6	X	4	X	4
	O_{32}	X	4	X	3	5	4
	O_{33}	X	X	11	X	9	13
4	O_{41}	9	X	7	9	X	6
	O_{42}	X	7	X	5	6	5
	O_{43}	2	3	4	X	X	4
5	O_{51}	X	4	5	3	X	4
	O_{52}	4	4	6	X	3	5
	O_{53}	3	4	X	5	6	X
6	O_{61}	X	3	7	4	5	X
	O_{62}	6	2	X	4	3	X
	O_{63}	5	4	3	X	X	4

reduce inventory and reduce costs has become a common concern of manufacturing companies. The JIT production mode emphasizes low inventory and high-quality production organization. In this mode, the delivery date of the order is the main performance index it considers, that is, the closer the order completion time is to the delivery date, the lower the product inventory, and the storage, The lower the handling cost. Therefore, as the performance index of the FJSP problem, the closer the completion time of the processing task is to the delivery date, the better the delivery date performance is. Generally, the maximum lead time index E_J and the maximum delay index T_J are used to measure the delivery performance index.

$$E_J = \max(d_J - C_J, 0) \tag{4.3}$$

d_J represents the delivery date of job J, and E_J is a non-negative value.

$$T_J = \max(C_J - d_J, 0) \tag{4.4}$$

Therefore, the optimization goal of the scheduling scheme is to minimize the maximum lead time E_J and the maximum delay T_J, which is:

$$f_2 = \min(E_{\max}) \tag{4.5}$$

$$f_3 = \min(T_{\max}) \tag{4.6}$$

In Eq. (4.5), $E_{\max} = \max(E_J)$. In Eq. (4.6), $T_{\max} = \max(T_J)$.

3. **Indicators based on machine tool load**

The load of the machine tool mainly refers to the working and running time of the machine tool. The machine tool load index reflects the utilization level of enterprise resources. According to the theory of constraints, the output of the bottleneck machine tool determines the final output of the enterprise. In order to improve production efficiency, the goal of scheduling should reduce the load of the bottleneck machine tool as much as possible, that is, minimize the maximum load of the machine tool. In order to judge the overall utilization rate of machine tools in the manufacturing workshop, the total load of all machine tools is also one of the evaluation indicators. In addition, to ensure balanced production, the overall load of the machine tool is required to be relatively balanced. Therefore, the load difference between the maximum load machine tool and the minimum load machine tool is also one of the evaluation indicators. Therefore, the load of the machine tool becomes an important index to measure the workshop scheduling, and the Eq. (4.7) expresses the load of the machine tool k.

$$L_k = \sum_{i=1}^{n} \sum_{j=1}^{p} T_{ijk} X_{ijk} \tag{4.7}$$

T_{ijk}: The processing time of the process O_{ij} of the job task J_j on the machine tool W_k;

X_{ijk}: Adjustment coefficient,

$$X_{ijk} = \begin{cases} 1 & When\ process\ O_{ij}\ is\ processed\ on\ machine\ tool\ W_k \\ 0 & When\ process\ O_{ij}\ is\ not\ processed\ on\ machine\ tool\ W_k \end{cases}$$

There are three main performance indicators for machine tool load as workshop scheduling:

(1) The maximum load machine load is minimized, that is:

$$f_4 = \min(L_{\max}) \tag{4.8}$$

In Eq. (4.8): $L_{\min} = \max(L_k)$.

(2) The total machine load is minimized, that is:

$$f_5 = \min\left(\sum_{k=1}^{m} L_k\right) \tag{4.9}$$

(3) The load difference between the maximum load machine tool and the minimum load machine tool is minimized, that is:

$$f_6 = \min(L_{\max} - L_{\min}) \tag{4.10}$$

In Eq. (4.10): $L_{\min} = \min(L_k)$.

4. **Indicators based on cost**

The production cost index directly reflects the impact of scheduling decisions on the economic benefits of enterprises. Production costs can be divided into constant costs and variable costs according to their nature. The constant cost has nothing to do with the number of processing tasks, including maintenance and depreciation fees for special machine tools and special process equipment, and related adjustment fees. Variable costs are related to the amount of processing tasks, including material costs, worker wages, and so on. The product cost incurred in the production process usually consists of the following parts:

Production cost: the production cost generated by time-consuming processing time, tool setting guide time, assembly and disassembly of parts, starting and stopping the machine, etc.

Late Penalty Cost: A one-time penalty and a penalty related to the length of the delay when the part is completed after the delivery date.

Storage cost: the storage and handling cost of semi-finished products waiting for processing, and the storage and handling cost that needs to be consumed when the product is completed ahead of schedule and cannot be delivered in advance.

Since the impact of lateness or early completion on scheduling decisions has been considered in the lead-time-based indicator, the cost-based indicator considers production costs and storage costs.

(1) Cost of production. Since the formation of the scheduling scheme is closely related to the time spent on processing tasks, in order to convert the production cost into a cost index, we uniformly divide the production cost into dynamic cost and static cost.

Dynamic cost: the total cost of resources consumed per unit time when the machine tools related to all job tasks are working. The cost can be expressed by Eq. (4.11):

$$C^v = \sum_{k=1}^{m} \left(\sum_{i=1}^{n} \sum_{j=1}^{p} T_{ijk} X_{ijk} \right) F_k^v \tag{4.11}$$

C^v: Dynamic cost;
F_k^v: Dynamic rates for machine tools W_k.
Other parameters in the formula are the same as Eq. (4.7)

Static cost: refers to the total cost that must be consumed per unit time when the equipment associated with the job is in the ready state. The cost can be expressed by Eq. (4.12).

$$C^s = \sum_{k=1}^{m}\left(T^* - \sum_{i=1}^{n}\sum_{j=1}^{p} T_{ijk}X_{ijk}\right)F_k^s \tag{4.12}$$

C^s: Static cost;
F_k^s: Static rates for machine tools W_k;
T^*: The completion time of the scheduling decision.
Other parameters in the formula are the same as Eq. (4.7)

Therefore, the scheduling decision index based on production cost is shown in Eq. (4.13)

$$f_7 = \min\left(C^p\right) = \min\left(C^v + C^s\right) \tag{4.13}$$

C^P: Total production cost of scheduling decision.

(2) Storage costs. There are three factors that affect storage costs: current value of parts, storage cost scaling factor, and storage time. The current value of the part is the value of the semi-finished product itself while waiting to be processed. According to the job task model, the processing process of parts is the process of increasing the value of raw materials. The value of parts increases after each operation, that is, the parts generate additional added value, that is, the value of parts increases with the progress of the processing process. The storage cost ratio coefficient refers to the percentage of the storage cost (including storage fee, occupied space fee, and opportunity cost of capital backlog, etc.) consumed by the semi-finished product when it is waiting for processing; the backlog time is from the beginning of waiting to the beginning Process the length of this period of time. Therefore, the storage cost can be expressed by Eq. (4.14):

$$C^w = \sum_{i=1}^{n}\left(\sum_{j=1}^{p}\left(C_{ij}^m + \sum_{j=1}^{p}\left(\sum_{k=1}^{m} T_{i(j-1)k}X_{i(j-1)k}F_k^v\right)\right)\eta_{ij}\left(ST_{ij} - \left(ST_{i(j-1)} + \sum_{k=1}^{m} T_{i(j-1)k}X_{i(j-1)k}\right)\right)\right) \tag{4.14}$$

C^w: Storage cost;
C_{ij}^m: The cost of raw materials for the job J_i;
η_{ij}: The ratio of the storage cost to the total value of the parts at this time when the job task J_i is processed to the O_{ij} process, referred to as the storage ratio coefficient;
ST_{ij}: The start time of the operation O_{ij} of the job J_i;
Other parameters in the formula are the same as Eq. (4.11)

According to the formula shown in Eq. (4.14), the storage cost-based scheduling decision evaluation index can be expressed by Eq. (4.15):

$$f_8 = \min\left(C^w\right) \tag{4.15}$$

4.2.2 Genetic Algorithm

GA is a local search algorithm that follows the evolution paradigm. Starting from an initial population, the algorithm applies genetic operators in order to produce offspring, which are presumably more fit than their ancestors. At each generation, every new individual corresponds to a solution, i.e., a schedule of the given FJSP instance. The strength of GA with respect to other local search algorithms is due to the fact that in a GA framework more strategies can be adopted together to find the right individuals, both in the initial population phase and in the dynamic generation phase. Then, a more variable search space can be explored at each algorithm step. The overall structure of the GAs can be described as follows:

(1) Gene coding: The genes of the chromosomes describe the assignment of operations to the machine tools, and the order that they appear in the chromosome describes the sequence of operations. Each chromosome represents a solution of the FJSP.
(2) Initializing population: Generally, the initial chromosomes are obtained randomly or in a way.
(3) Copy, crossover and mutation: At each iteration, the best chromosomes are chosen for reproduction. The new generation is obtained by changing the assignment of the operations to the machines (assignment crossover, assignment mutation) and by changing the sequencing of operations.
(4) Stop criterion: Fixed number of generations is reached. If the stop criterion is satisfied, the algorithm ends and the best chromosome, together with the corresponding schedule, are given as output. Otherwise, the algorithm iterates again the steps above.

4.2.3 Scheduling Algorithm Based on GA

The FJSP in Table 4.1 is taken as a sample to illustrate the application of GA.

1. **Gene coding**

Gene coding is the first and crucial step for GA to solve the FJSP. By transforming the FJSP into a reasonable and efficient chromosome expression scheme, it has an important impact on further genetic manipulation. As mentioned in the previous section, the essence of the FJSP is to solve two problems. One is to select a suitable machine tool for each operation of the processing job, and the other is to determine the sequence arrangement and start time of each operation of the processing job. For the above two sub-problems, when solving the FJSP, there are mainly two ways of gene coding: integrated and segmented. In the integrated coding method, each gene (h, j, i) in the chromosome represents a process job, which indicates that the hth operation of part j is processed on machine tool i. In the segmented coding method, the chromosome is divided into two parts, one part represents the machine tool selected by the operation, and the other part represents the sequence arrangement of the operation. Since the problem of expressing FJSP in the segmented coding method is intuitive and clear, and the genetic operation is easy to design, in recent years, there have been many studies on the segmented genetic coding method.

So, the chromosome structure is designed according to the segmented gene coding method, and the operations of chromosome population initialization, selection, crossover, and mutation are improved. Chromosomes are made up of two parts:

Genetic code of machine tools: This part of the genetic code represents the number of the processing machine tool for each operation in each job, and is in one-to-one correspondence with the operations in the genetic code of the operations.

Genetic code of Operations: This part of the gene code is composed of job numbers, which respectively represent the first operation, the second operation, the nth operation according to the order of job numbers.

For example, the chromosome A shown in Table 4.2 is a reasonable chromosome generated for the 6X6 flexible job scheduling problem shown in Table 4.1.

In Table 4.2, the first gene of Operation Gene in chromosome A is 1, which represents the operation O_{11}. The corresponding Machine Gene is 1, indicating that the operation O_{11} is processed on the machine tool W_1, which is the first sequence bit in the activity scheduling scheme. The 18th gene 3 of Operation Gene represents the operation O_{33}. The corresponding gene 5 of the machine tool indicates that the selected machine tool is W_5, which is the 18th sequence bit in the activity scheduling scheme.

The scheduling scheme represented by the above chromosomes can be expressed by Formula (4.16).

$$S = \{O_{111}, O_{215}, O_{314}, O_{416}, O_{514}, O_{612}, O_{426}, O_{125}, O_{525}, O_{224}, O_{131}, O_{431}, O_{622}, O_{531}, O_{233}, O_{633}, O_{324}, O_{335}\} \tag{4.16}$$

The Gantt chart of the scheduling scheme is shown in Fig. 4.4.

Through the above-mentioned chromosome expression scheme, it is possible not only to represent the processing machine tools of each operation, but also to determine the sequence arrangement of each operation. However, the starting time of each operation on the corresponding machine tool needs to be determined separately.

2. **Initializing population**

The initial population of chromosomes contains the original chromosomes for genetic manipulation, and the size of the initial population is determined by the scale of the FJSP. Most of the literature adopts the random initialization method for the population

Table 4.2 Chromosome A

Operation gene	1	2	3	4	5	6	4	1	5	2	1	4	6	5	2	6	3	3
Operation	O_{11}	O_{21}	O_{31}	O_{41}	O_{51}	O_{61}	O_{42}	O_{12}	O_{52}	O_{22}	O_{13}	O_{43}	O_{62}	O_{53}	O_{23}	O_{63}	O_{32}	O_{33}
Machine gene	1	5	4	6	4	2	6	5	5	4	1	1	2	1	3	3	4	5

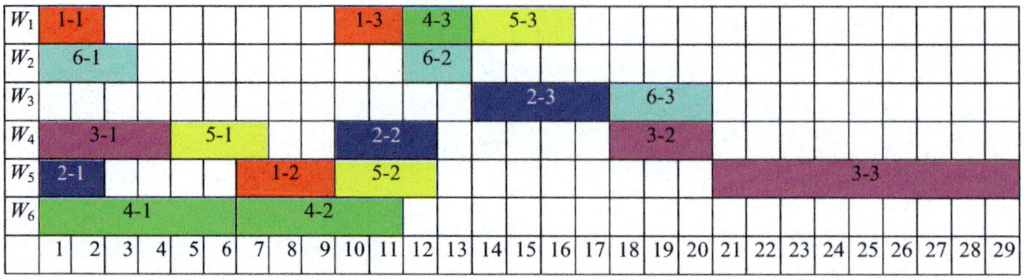

Fig. 4.4 Gantt chart of the scheduling scheme of chromosome A

initialization of the FJSP. The advantage of this method is that it is simple and clear, but it often causes various difficulties. For example, due to the large search space of the initial population, the convergence speed of the algorithm is slowed, and the load of the processing machine tool is unbalanced, and so on.

The population initialization method in this section builds on the job schedule. Considering the load of the machine tool and the diversity of the population, the initialization population adopts two methods, namely, GMT and RS. GMT is based on the least processing time on global. RS is based on random selection of operations and machine tools.

The workflow of the GMT method is as follows:

Step 1: Select the process O_{ij} with the smallest processing time and its machine tool W_k in the range of all operations and machine tools, and record its processing time T_{ijk}.

Step 2: In the machine tool load table, the processing time of all processes of the corresponding machine tool increases accordingly, that is, $T'_{dfk} = T_{dfk} + T_{ijk}$.

Step 3: Repeat the above steps until the processing machines for all processes are determined and their processing time is recorded.

Take the data shown in Table 4.1 as an example to illustrate the above process. The simple calculation of determining the gene part of the machine tool by the GMT method is shown in Table 4.3.

In Table 4.3, the operation and the machine tool with the minimum processing time are searched in the global scope. The processing time of O_{11} on machine tools W_5 and W_1 is 2 s, the processing time of O_{13} on machine tool W_1 is 2 s, the processing time of O_{21} on machine tool W_5 is 2 s, the processing time of O_{43} on machine tool W_1 is 2 s, and the processing time of O_{62} on machine tool W_2 is 2 s, and the above operations and machine tools meet the requirements. If O_{11} is randomly selected to be processed on machine tool W_5, the processing time of all operations on W_5 except O_{11} will be increased by the processing time of O_{11} on machine tool W_5, which is 2 s, as shown as bold numbers in the second group of Table 4.3. Similarly, continue to search for the operation with the smallest load or processing time in the global scope. The processing time of O_{13} on machine tool W_1 is 2 s, and the processing time of O_{62} on machine tool W_2 is 2 s. Both of the above two operations meet the requirements, so select randomly the operation O_{13}, which is processed on machine tool W_1, the processing time of all operations on machine tool W_1 except O_{13} is

Table 4.3 The simple calculation of GMT method

	W_1	W_2	W_3	W_4	W_5	W_6	W_1	W_2	W_3	W_4	W_5	W_6	W_1	W_2	W_3	W_4	W_5	W_6
O_{11}	2	3	X	5	**2**	3	2	3	X	5	**2**	3	**4**	3	X	5	2	3
O_{12}	4	3	5	X	3	X	4	3	5	X	**5**	X	**6**	3	5	X	5	X
O_{13}	2	X	5	4	X	4	**2**	X	5	4	**X**	4	**2**	X	5	4	X	4
O_{21}	3	X	5	3	2	X	3	X	5	3	**4**	X	**5**	X	5	3	4	X
O_{22}	4	X	3	3	X	5	4	X	3	3	**X**	5	**6**	X	3	3	X	5
O_{23}	X	X	4	5	7	9	X	X	4	5	**9**	9	**X**	X	4	5	9	9
O_{31}	5	6	X	4	X	4	5	6	X	4	**X**	4	**7**	6	X	4	X	4
O_{32}	X	4	X	3	5	4	X	4	X	3	**7**	4	**X**	4	X	3	7	4
O_{33}	X	X	11	X	9	13	X	X	11	X	**11**	13	**X**	X	11	X	11	13
O_{41}	9	X	7	9	X	6	9	X	7	9	**X**	6	**11**	X	7	9	X	6
O_{42}	X	7	X	5	6	5	X	7	X	5	**8**	5	**X**	7	X	5	8	5
O_{43}	2	3	4	X	X	4	2	3	4	X	**X**	4	**4**	3	4	X	X	4
O_{51}	X	4	5	3	X	4	X	4	5	3	**X**	4	**X**	4	5	3	X	4
O_{52}	4	4	6	X	3	5	4	4	6	X	**5**	5	**6**	4	6	X	5	5
O_{53}	3	4	X	5	6	X	3	4	X	5	**8**	X	**5**	4	X	5	8	X
O_{61}	X	3	7	4	5	X	X	3	7	4	**7**	X	**X**	3	7	4	7	X
O_{62}	6	2	X	4	3	X	6	2	X	4	**5**	X	**8**	**2**	X	4	5	X
O_{63}	5	4	3	X	X	4	5	4	3	X	**X**	4	**7**	4	3	X	X	4
	W_1	W_2	W_3	W_4	W_5	W_6	W_1	W_2	W_3	W_4	W_5	W_6	W_1	W_2	W_3	W_4	W_5	W_6
O_{11}	4	**5**	X	5	2	3	4	5	X	**8**	2	3	4	5	**X**	8	2	3
O_{12}	6	**5**	5	X	5	X	6	5	5	**X**	5	X	6	5	**8**	X	5	X
O_{13}	2	**X**	5	4	X	4	2	X	5	**7**	X	4	2	X	**8**	7	X	4
O_{21}	5	**X**	5	**3**	4	X	5	X	5	**3**	4	X	5	X	**8**	3	4	X
O_{22}	6	**X**	3	3	X	5	6	X	3	**6**	X	5	6	X	**6**	6	X	5
O_{23}	X	**X**	4	5	9	9	X	X	4	**8**	9	9	X	X	**7**	8	9	9
O_{31}	7	**8**	X	4	X	4	7	8	X	**7**	X	4	7	8	**X**	7	X	4
O_{32}	X	**6**	X	3	7	4	X	6	X	**6**	7	4	X	6	**X**	6	7	4
O_{33}	X	**X**	11	X	11	13	X	X	11	**X**	11	13	X	X	**14**	X	11	13
O_{41}	11	**X**	7	9	X	6	11	X	7	**12**	X	6	11	X	**10**	12	X	6
O_{42}	X	**9**	X	5	8	5	X	9	X	**8**	8	5	X	9	**X**	8	8	5
O_{43}	4	**5**	4	X	X	4	4	5	4	**X**	X	4	4	5	**7**	X	X	4
O_{51}	X	**6**	5	3	X	4	X	6	5	**6**	X	4	X	6	**8**	6	X	4
O_{52}	6	**6**	6	X	5	5	6	6	6	**X**	5	5	6	6	**9**	X	5	5
O_{53}	5	**6**	X	5	8	X	5	6	X	**8**	8	X	5	6	**X**	8	8	X
O_{61}	X	**5**	7	4	7	X	X	5	7	**7**	7	X	X	5	**10**	7	7	X
O_{62}	8	**2**	X	4	5	X	8	2	X	**7**	5	X	8	2	**X**	7	5	X
O_{63}	7	**6**	3	X	X	4	7	6	**3**	**X**	X	4	7	6	**3**	X	X	4

Table 4.4 The operations and machine tools determined by GMT

Job	Operation	Machine					
		W_1	W_2	W_3	W_4	W_5	W_6
1	O_{11}					2	
	O_{12}		3				
	O_{13}	2					
2	O_{21}				3		
	O_{22}			3			
	O_{23}			4			
3	O_{31}						4
	O_{32}				3		
	O_{33}						13
4	O_{41}				9		
	O_{42}					6	
	O_{43}	2					
5	O_{51}						4
	O_{52}					3	
	O_{53}	3					
6	O_{61}		3				
	O_{62}		2				
	O_{63}			3			

increased by the processing time of O_{13} on machine tool W_1, which is 2 s, as shown as bold numbers in the third group of Table 4.3. Perform the above operations in a loop until the processing machines for all operations are determined. See Appendix 1 for the detailed calculation process.

The machine tool for each operation can be determined by the GMT method, as shown in Table 4.4.

IF the order of the operations is randomly arranged, the chromosomes B obtained are shown in Table 4.5.

The workflow of the RS method is as follows:

Step 1: Select the machine tool W_k, on which the first operation O_{ij} in the sequence of operations is processed with the minimum processing time, and record its processing time T_{ijk}.

Step 2: In the machine tool load table, the processing time of all operations on the corresponding machine tool behind the operation O_{ij} increases accordingly, that is, $T'_{dfk} = T_{dfk} + T_{ijk}$.

Step 3: Select the machine tool W'_k, on which the first operation O'_{ij} in the sequence of operations is processed with the minimum processing time, and record its processing time T'_{ijk}.

Table 4.5 Chromosome *B*

Operation Gene	1	2	3	4	5	6	3	5	2	6	1	3	4	5	1	6	4	2
Operation	O_{11}	O_{21}	O_{31}	O_{41}	O_{51}	O_{61}	O_{32}	O_{52}	O_{22}	O_{62}	O_{12}	O_{33}	O_{42}	O_{53}	O_{13}	O_{63}	O_{43}	O_{23}
Machine gene	5	4	6	4	6	2	4	5	3	2	2	6	5	1	1	3	1	3

Step 4: In the machine tool load table, the processing time of all operations on the corresponding machine tool behind the operation O_{ij} increases accordingly, that is, $T'_{dfk} = T_{dfk} + T'_{ijk}$.

Step 5: Repeat the above steps until the processing machines for all operations are determined, and their processing time is recorded.

Randomly sort the sequence of operations for the 6X6 FJSP shown in Table 4.1. Because the related operations of the same job must be processed in sequence, for example, the processing of operation O_{11} must be performed before operation O_{12} and operation O_{13}. So randomly sort the sequence of operations should meet the above requirements. Take the data shown in Table 4.6 as an example to illustrate the above process.

In Table 4.6, search for the machine tools, on which the first operation O_{11} is processed with the minimum time, then W_1 and W_5 is found. In this example, W_5 is selected. The process processing time of all operations on W_5 except the first operation O_{11} increases the time of the first operation O_{11} on W_5, which is 1 s, as shown as bold numbers in the second group of Table 4.6. In the same way, search for the machine tools, on which the second operation O_{21} is processed with the minimum time. The processing time of O_{21} on W_1 and W_5 both is 3 s. In this example, W_4 is selected. The process processing time of all operations on W_4 except the second operation O_{21} increases the time of the second operation O_{21} on W_4, which is 3 s, as shown as bold numbers in the third group of Table 4.6. Perform the above operations line by line until the processing machines for all operations are determined. See Appendix 2 for the detailed calculation process.

The machine tool for each operation can be determined by the RS method, as shown in Table 4.7.

IF the order of the operations is randomly arranged, the chromosomes C obtained are shown in Table 4.8.

3. **Selection, Cross and Mutation**

The minimization of the maximum completion time $C_{\max}$ is taken as the optimization goal to solve the FJSP. Accordingly, we determine the fitness function as shown in Eq. (4.17):

$$F(x) = \mathrm{Max}(C_{\max}) - C_{\max} \tag{4.17}$$

The selection of individuals to produce successive generations plays an important role in GAs. A probabilistic selection is performed based upon the individual's fitness such that the better individuals could be selected. In order to guarantee the astringency of GAs, the optimal individual in one generation must be retained to the next generation directly. Other individuals in population are selected using the selection criterion. Generally, the selection criterion includes two well-known methods: tournament and linear ranking. In this paper, the criterion of tournament is adopted which has to be repeated until the number of individuals equals the population size. In order to avoid local optimum, the probability of selection is set to be 5–15%.

Table 4.6 The simple calculation of RS method

	W_1	W_2	W_3	W_4	W_5	W_6	W_1	W_2	W_3	W_4	W_5	W_6	W_1	W_2	W_3	W_4	W_5	W_6
O_{11}	2	3	X	5	**2**	3	2	3	X	5	**2**	3	2	3	X	5	2	3
O_{21}	3	X	5	3	2	X	3	X	5	**3**	**4**	X	3	X	5	**3**	4	X
O_{31}	5	6	X	4	X	4	5	6	X	4	**X**	4	5	6	X	**7**	X	**4**
O_{41}	9	X	7	9	X	6	9	X	7	9	**X**	6	9	X	7	**12**	X	6
O_{51}	X	4	5	3	X	4	X	4	5	3	**X**	4	X	4	5	**6**	X	4
O_{61}	X	3	7	4	5	X	X	3	7	4	**7**	X	X	3	7	**7**	7	X
O_{42}	X	7	X	5	6	5	X	7	X	5	**8**	5	X	7	X	**8**	8	5
O_{12}	4	3	5	X	3	X	4	3	5	X	**5**	X	4	3	5	**X**	5	X
O_{52}	4	4	6	X	3	5	4	4	6	X	**5**	5	4	4	6	**X**	5	5
O_{22}	4	X	3	3	X	5	4	X	3	3	**X**	5	4	X	3	**6**	X	5
O_{13}	2	X	5	4	X	4	2	X	5	4	**X**	4	2	X	5	**7**	X	4
O_{43}	2	3	4	X	X	4	2	3	4	X	**X**	4	2	3	4	**X**	X	4
O_{62}	6	2	X	4	3	X	6	2	X	4	**5**	X	6	2	X	**7**	5	X
O_{53}	3	4	X	5	6	X	3	4	X	5	**8**	X	3	4	X	**8**	8	X
O_{23}	X	X	4	5	7	9	X	X	4	5	**9**	9	X	X	4	**8**	9	9
O_{63}	5	4	3	X	X	4	5	4	3	X	**X**	4	5	4	3	**X**	X	4
O_{32}	X	4	X	3	5	4	X	4	X	3	**7**	4	X	4	X	**6**	7	4
O_{33}	X	X	11	X	9	13	X	X	11	X	**11**	13	X	X	11	**X**	11	13
	W_1	W_2	W_3	W_4	W_5	W_6	W_1	W_2	W_3	W_4	W_5	W_6	W_1	W_2	W_3	W_4	W_5	W_6
O_{11}	2	3	X	5	2	3	2	3	X	5	2	3	2	3	X	5	2	3
O_{21}	3	X	5	3	4	X	3	X	5	3	4	X	3	X	5	3	4	X
O_{31}	5	6	X	7	X	**4**	5	6	X	7	X	4	5	6	X	7	X	4
O_{41}	9	X	**7**	12	X	**10**	9	X	**7**	12	X	10	9	X	7	12	X	6
O_{51}	X	4	5	6	X	**8**	X	**4**	**12**	6	X	8	X	**4**	5	6	X	4
O_{61}	X	3	7	7	7	**X**	X	3	**14**	7	7	X	X	**7**	7	7	7	X
O_{42}	X	7	X	8	8	**9**	X	7	**X**	8	8	9	X	**11**	X	8	8	5
O_{12}	4	3	5	X	5	**X**	4	3	**12**	X	5	X	4	**7**	5	X	5	X
O_{52}	4	4	6	X	5	**9**	4	4	**13**	X	5	9	4	**8**	6	X	5	5
O_{22}	4	X	3	6	X	**9**	4	X	**10**	6	X	9	4	**X**	3	6	X	5
O_{13}	2	X	5	7	X	**8**	2	X	**12**	7	X	8	2	**X**	5	7	X	4
O_{43}	2	3	4	X	X	**8**	2	3	**11**	X	X	8	2	**7**	4	X	X	4
O_{62}	6	2	X	7	5	**X**	6	2	**X**	7	5	X	6	**6**	X	7	5	X
O_{53}	3	4	X	8	8	**X**	3	4	**X**	8	8	X	3	**8**	X	8	8	X
O_{23}	X	X	4	8	9	**13**	X	X	**11**	8	9	13	X	**X**	4	8	9	9
O_{63}	5	4	3	X	X	**8**	5	4	**10**	X	X	8	5	**8**	3	X	X	4
O_{32}	X	4	X	6	7	**8**	X	4	**X**	6	7	8	X	**8**	X	6	7	4
O_{33}	X	X	11	X	11	**17**	X	X	**18**	X	11	17	X	X	11	X	11	13

Table 4.7 The operations and machine tools determined by RS

Job	Operation	Machine					
		W_1	W_2	W_3	W_4	W_5	W_6
1	O_{11}					2	
	O_{12}	4					
	O_{13}			5			
2	O_{21}				3		
	O_{22}						5
	O_{23}				5		
3	O_{31}						4
	O_{32}						4
	O_{33}					9	
4	O_{41}			7			
	O_{42}				5		
	O_{43}		3				
5	O_{51}		4				
	O_{52}	4					
	O_{53}		4				
6	O_{61}					5	
	O_{62}					3	
	O_{63}	5					

The chromosome is divided into two parts in this segmented gene coding method, namely, the machine tool gene code and the operation gene code. Therefore, the crossover operation is performed on these two parts.

The gene part of the operation adopts a two-point crossover operation, and the specific algorithm flow is as follows:

Step 1: For all chromosomes in the population, judge whether to perform crossover operation with the crossover probability, and determine the two chromosomes B and C for crossover operation.

Step 2: Randomly generate two intersection points p, q.

Step 3: Take out the genes between the two intersection points p and q in the operation gene part of the chromosome B, and keep the genes outside the intersection points unchanged.

Step 4: Find the missed genes in the operation gene part of another parent chromosome C. These missed genes are inserted into the position between the two intersection points of B according to the original arrangement order of C. A chromosome D forms with the feathers of the chromosome B and chromosome C.

Step 5: Fill in the machine tool selected by the corresponding operations in parent chromosome B to the corresponding operation position between the intersection points of offspring chromosome D.

Table 4.8 Chromosome C

Operation gene	1	2	3	4	5	6	4	1	5	2	1	4	6	5	2	6	3	3
Operation	O_{11}	O_{21}	O_{31}	O_{41}	O_{51}	O_{61}	O_{42}	O_{12}	O_{52}	O_{22}	O_{13}	O_{43}	O_{62}	O_{53}	O_{23}	O_{63}	O_{32}	O_{33}
Machine gene	5	4	6	3	2	5	4	1	1	6	3	2	5	2	4	1	6	5

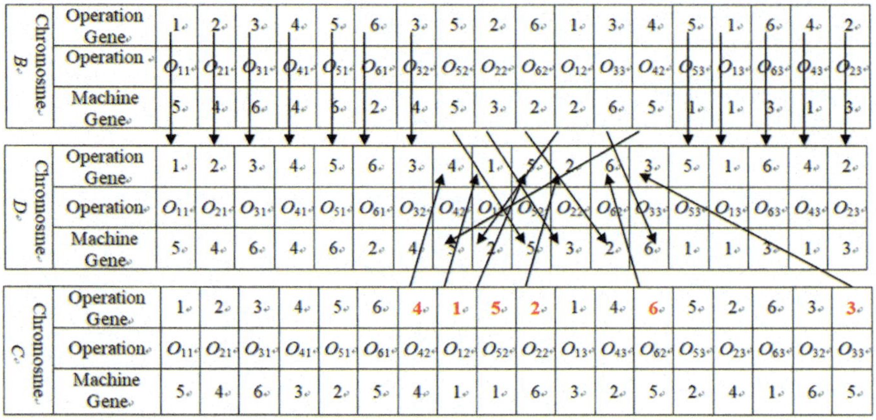

Fig. 4.5 Two-point crossover algorithm for operation gene part

Take chromosome B in Table 4.5 and chromosome C in Table 4.8 as examples to illustrate the process of the two-point crossover algorithm. The intersection points of chromosome B are taken as two genes, 8 and 13, and a new offspring is formed after crossover operation. Chromosome D, the algorithm execution process is shown in Fig. 4.5, and chromosome D is shown in Table 4.9.

Similarly, the intersection point of chromosome C is 8 and 13, and after the crossover operation, a new offspring chromosome E is formed, as shown in Table 4.10.

The genetic part of the machine tool adopts the uniform crossover algorithm. Due to the limitation of the processing machine, two-point crossover may cause the failure of the crossover operation of the machine tool genetic part. Therefore, a uniform crossover operation is adopted in the machine tool gene part. The crossover algorithm flow is as follows:

Step 1: Select the above chromosomes D and E generated by performing the crossover operation on the operation gene part.

Step 2: Randomly generate two intersection points p, q, which represent the job numbers. The crossover will be performed on the corresponding jobs.

Step 3: Keep the operation gene part unchanged, cross the machine tools limited by all operations of job p and q in chromosome D and the machine tools limited by all operations of job p and q in chromosome E, and replace the machine gene part of the corresponding operations.

The chromosome D in Table 4.9 and the chromosome E in Table 4.10 are taken to perform crossover operation. For the chromosome D, the crossover positions are 3 and 4. The algorithm execution process is shown in Fig. 4.6. Chromosome F is obtained as shown in Table 4.11. For the chromosome D, the crossover positions are 2 and 5. After the crossover operation, a new chromosome G is formed, as shown in Table 4.12.

Table 4.9 The chromosome D

Operation gene	1	2	3	4	5	6	3	4	1	5	2	6	3	5	1	6	4	2
Operation	O_{11}	O_{21}	O_{31}	O_{41}	O_{51}	O_{61}	O_{32}	O_{42}	O_{12}	O_{52}	O_{22}	O_{62}	O_{33}	O_{53}	O_{13}	O_{63}	O_{43}	O_{23}
Machine gene	5	4	6	4	6	2	4	5	2	5	3	2	6	1	1	3	1	3

Table 4.10 The chromosome E

Operation gene	1	2	3	4	5	6	4	5	2	1	4	1	6	5	2	6	3	3
Operation	O_{11}	O_{21}	O_{31}	O_{41}	O_{51}	O_{61}	O_{42}	O_{52}	O_{22}	O_{12}	O_{43}	O_{13}	O_{62}	O_{53}	O_{23}	O_{63}	O_{32}	O_{33}
Machine gene	5	4	6	3	2	5	4	1	6	1	2	3	5	2	4	1	6	5

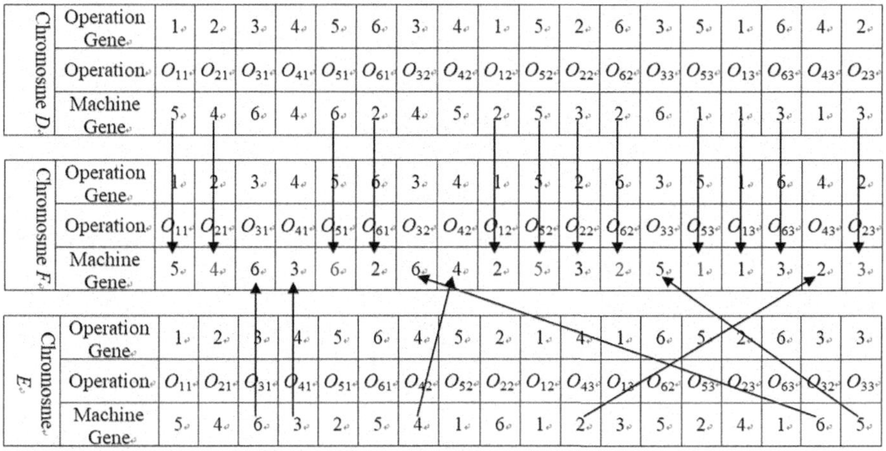

Fig. 4.6 Uniform crossover algorithm for machine tool gene part

The main purpose of the mutation operation is to make the genetic algorithm have local random search ability, and maintain population diversity to prevent premature convergence. When the genetic algorithm is close to the optimal solution neighborhood through the crossover operation, the local random search ability of the mutation operation can be used to accelerate the convergence to the optimal solution. Obviously, the mutation probability in this case should take a smaller value. Due to the adoption of the segmented genetic coding method, the mutation operation also performs different genetic operations on the two parts of the genetic code. The workflow for the mutation operation on the operation genetic part is as follows:

Step 1: All chromosomes in the population are judged whether to perform mutation operation according to the mutation probability set in advance, and two chromosomes B and C for mutation operation are selected to perform mutation operation.

Step 2: Randomly generate two mutation positions p, q, and exchange the genes at the two positions p and q.

Step 3: Check whether the operation gene part in the exchange position meets the requirements, that is, the latter operation can only be carried out after the previous operation is completed, if the requirements are not met, return to Step 2.

Step 4: Exchange the machine tools limited by the operation of the two exchanged positions p and q.

For the chromosome B in Table 4.5, the mutation points are taken as 6 and 12. After testing, if the operation genes of 6 and 12 are exchanged, it will cause operation O_{33} to be arranged before operation O_{32}. Obviously, this does not meet the actual process constraints. Therefore, chromosome B performs mutation by taking the mutation points 4 and 9, and the chromosome H obtained is shown in Table 4.13.

Similarly, for chromosome C in Table 4.8, the mutation points are taken as 6 and 12. After testing, if the operation genes of 6 and 12 are exchanged, it will cause operation O_{43} to be arranged before operation O_{42}, which does not meet the

Table 4.11 The chromosome F

Operation gene	1	2	3	4	5	6	3	4	1	5	2	6	3	5	1	6	4	2
Operation	O_{11}	O_{21}	O_{31}	O_{41}	O_{51}	O_{61}	O_{32}	O_{42}	O_{12}	O_{52}	O_{22}	O_{62}	O_{33}	O_{53}	O_{13}	O_{63}	O_{43}	O_{23}
Machine gene	5	4	6	3	6	2	6	4	2	5	3	2	5	1	1	3	2	3

Table 4.12 The chromosome G

Operation gene	1	2	3	4	5	6	4	5	2	1	4	1	6	5	2	6	3	3
Operation	O_{11}	O_{21}	O_{31}	O_{41}	O_{51}	O_{61}	O_{42}	O_{52}	O_{22}	O_{12}	O_{43}	O_{13}	O_{62}	O_{53}	O_{23}	O_{63}	O_{32}	O_{33}
Machine gene	5	4	6	3	6	5	4	5	3	1	2	3	5	1	3	1	6	5

Table 4.13 The chromosome H

Operation gene	1	2	3	2	5	6	3	5	4	6	1	3	4	5	1	6	4	2
Operation	O_{11}	O_{21}	O_{31}	O_{22}	O_{51}	O_{61}	O_{32}	O_{52}	O_{41}	O_{62}	O_{12}	O_{33}	O_{42}	O_{53}	O_{13}	O_{63}	O_{43}	O_{23}
Machine gene	5	4	6	3	6	2	4	5	4	2	2	6	5	1	1	3	1	3

actual process constraints. Therefore, chromosome C performs mutation by taking the mutation points 6 and 9, and the obtained chromosome I is shown in Table 4.14.

The workflow for performing mutation operations on the machine tool gene part is as follows:

Step 1: Randomly select the above chromosomes H and I generated by performing the mutation on the operation gene part.

Step 2: Randomly generate two mutation points p, q, which represent the number of the machine tool, on which the mutation is performed.

Step 3: Judging whether all operations processed on machine tool p can be processed on machine tool q, if the requirements are not met, return to Step 2.

Step 4: Judging whether all operations processed on machine tool q can be processed on machine tool p, if the requirements are not met, return to Step 2.

Step 5: Keep the operation gene part unchanged, and exchange machine tool p and machine tool q.

For chromosome H in Table 4.13, the mutation points are taken as 2 and 5. A new offspring chromosome J is shown in Table 4.15 after mutation.

For chromosome I in Table 4.14, the mutation points are taken as 1 and 3. A new offspring chromosome K is shown in Table 4.16 after mutation.

4. **Gene decoding**

The chromosome is represented as a concrete scheduling scheme by gene decoding. As shown in Fig. 4.4, the scheduling is expressed as a specific Gantt chart. According to the above-mentioned genetic coding method and the process of solving the optimal scheduling scheme through genetic algorithm, it can be determined that the most critical issue in genetic decoding is to determine the start time of each operation. Since the machine tool gene part has determined the machine tool for each operation, the processing time of this operation has also been determined. The operation gene part only determines the sequence of starting each operation, but does not determine the specific start time of each operation. To determine the start time of each operation, two time nodes need to be considered, that is, the completion time of the previous operation of the current job and the completion time of the last operation of the machine tool selected for this operation. Therefore, the start time ST_{ijk} of the operation O_{ij} on machine tool W_k can be expressed by Eq. (4.18).

$$ST_{ijk} = \max\left(ST_{ij-1} + T_{ij-1}, SE_k\right) \tag{4.18}$$

In Eq. (4.18), ST_{ij} is the start time of the operation O_{ij} of the job J_i, where $ST_{i0} = 0$, $T_{i0} = 0$.

SE_k is the completion time of the last operation of the machine tool W_k, which can be expressed by Eq. (4.19).

$$SE_k = \begin{cases} 0 & L_k = 0 \\ ST_{ij-1k} + T_{ij-1k}X_{ij-1k} & L_k \neq 0 \end{cases} \tag{4.19}$$

Table 4.14 The chromosome I

Operation gene	1	2	3	4	5	5	4	1	6	2	1	4	6	5	2	6	3	3
Operation	O_{11}	O_{21}	O_{31}	O_{41}	O_{51}	O_{52}	O_{42}	O_{12}	O_{61}	O_{22}	O_{13}	O_{43}	O_{62}	O_{53}	O_{23}	O_{63}	O_{32}	O_{33}
Machine gene	5	4	6	3	2	1	4	1	5	6	3	2	5	2	4	1	6	5

Table 4.15 The chromosome J

Operation Gene	1	2	3	2	5	6	3	5	4	6	1	3	4	5	1	6	4	2
Operation	O_{11}	O_{21}	O_{31}	O_{22}	O_{51}	O_{61}	O_{32}	O_{52}	O_{41}	O_{62}	O_{12}	O_{33}	O_{42}	O_{53}	O_{13}	O_{63}	O_{43}	O_{23}
Machine gene	2	4	6	3	6	5	4	2	4	5	5	6	2	1	1	3	1	3

Table 4.16 The chromosome K

Operation gene	1	2	3	4	5	5	4	1	6	2	1	4	6	5	2	6	3	3
Operation	O_{11}	O_{21}	O_{31}	O_{41}	O_{51}	O_{52}	O_{42}	O_{12}	O_{61}	O_{22}	O_{13}	O_{43}	O_{62}	O_{53}	O_{23}	O_{63}	O_{32}	O_{33}
Machine gene	5	4	6	1	2	3	4	3	5	6	1	2	5	2	4	3	6	5

In Eq. (4.19), L_k is the load of the machine tool W_k. $X_{ij\text{-}1k}$ is adjustment coefficient.

$$X_{ij-1k} = \begin{cases} 1 & O_{ij}\ is\ performed\ on\ machine\ tool\ W_k \\ 0 & O_{ij}\ is\ not\ performed\ on\ machine\ tool\ W_k \end{cases} \tag{4.20}$$

The chromosome I in Table 4.14 is taken as an example to illustrate the calculation process of the start time of each operation, as shown in Table 4.17, and based on this, generate the Gantt chart of the chromosome, as shown in Fig. 4.7.

Table 4.17 Decoded table of the chromosome I

Operation	T_{ij}	$T_{ij.1}$	$ST_{ij.1}$	SE_k	ST_{ijk}					
					W_1	W_2	W_3	W_4	W_5	W_6
O_{11}	2	0	0	0					0	
O_{21}	3	0	0	0				0		
O_{31}	4	0	0	0						0
O_{41}	7	0	0	0			0			
O_{51}	4	0	0	0		0				
O_{52}	4	4	0	0	4					
O_{42}	5	7	0	3				7		
O_{12}	4	3	0	8	8					
O_{61}	5	0	0	2					2	
O_{22}	5	3	0	4						4
O_{13}	5	4	8	7			12			
O_{43}	3	5	7	0		12				
O_{62}	3	5	2	7					7	
O_{53}	4	4	4	15		15				
O_{23}	5	5	7	9				12		
O_{63}	5	3	7	12	12					
O_{32}	4	4	0	9						9
O_{33}	9	3	9	10					13	

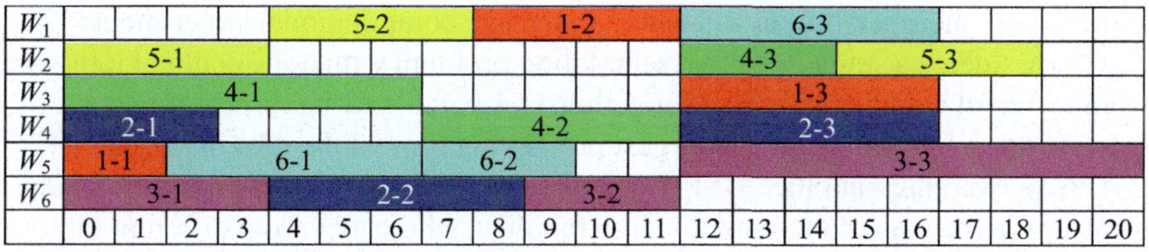

Fig. 4.7 The Gantt chart of the chromosome I

4.3 Applications of Ant Colony Algorithms for Scheduling Parallel Batch Processing Machines

4.3.1 Introduction

Simulation is important to model and analyze the actual performance of production systems. An excellent simulation framework consists of simulation software, expert system and artificial intelligence techniques, which controls the unexpected disturbances to an acceptable degree of production systems and provides the reasonable means to balance and collaborate the behaviour of production systems.

As compared to other production systems such as mass production and job shop production, batch production system plays a major role in modern manufacturing systems because of rapid variation of customer requirements, Scheduling the parallel batch processing machines reasonably is increasingly important to improve efficiency of production systems. This paper studies parallel processing machines scheduling problem that Dobson and Nambimadom called the batch loading and scheduling problem [21]. The batch processing machine scheduling problems in many literatures are decomposed into two different subproblems based on the characteristics of the batch processing system under consideration such as: (1) batch processing time, (2) family-based scheduling, (3) job sizes. In the scheduling, all the jobs can be partitioned into a number of families where jobs from the same family have the same processing time. The jobs from different families cannot be processed together. The processing time of a batch may be dependent on the constituent jobs of that batch (Webster and Baker) or may be constant [22]. As scheduling job shop, the objectives of scheduling parallel batch processing machines include $C_{\max}$, $L_{\max}$, $\sum C_i$, $\sum \omega_i C_i$, and so on [23–25, 34].

Lee et al., the first researchers who popularized scheduling burn-in operations, gave efficient algorithms for minimizing $L_{\max}$ and $\sum C_i$, assuming reasonable processing times and due dates [24]. Lee and Uzsoy studied the problem of minimizing $C_{\max}$ on a single batch processing machine in the presence of dynamic job arrivals. Scheduling a single batch processing machine where boards are also considered as a secondary resource constraint has been considered by Kempf et al. [25]. The polynomial and pseudo-polynomial time algorithms for several special cases is proposed in the paper. They also develop efficient heuristics for the general problem and evaluate their performance through extensive computational experiments [26].

Uzsoy studies a single machine scheduling problem with non-identical jobs sizes from different families processed together in a batch and the processing time of a batch is given by computing the largest processing time of the jobs in the batch [27].

Uzsoy examines another BSP to optimize a number of different performance measures in the case of incompatible job families, where jobs from different families cannot be processed together in the same batch, all jobs of the same family have a common processing time, and each job requires one unit of machine capacity [28]. Mehta and Uzsoy develop a DP algorithm with a polynomial time complexity in case of fixed number of incompatible job families and fixed batch machine capacity and also propose heuristic procedures such as batched apparent tardiness cost rule and decomposition heuristic to get near optimal solutions in practical computational time [29]. Devpura et al. propose six heuristics based on the decomposition approach for the scheduling single batch processor to minimize the total weighted tardiness objective with incompatible job families. The batch formation problem is solved using dispatching rules [30]. Perez et al. experiment with the six proposed heuristics in Devpura et al. for a large set of randomly generated problem instances as in Mehta and Uzsoy and give the detailed interpretation of the results [31].

Wang and Uzsoy coupled a random key-based representation genetic algorithm (GA) with a dynamic programming algorithm to minimize $L_{\max}$ on a batch processing machine in the presence of job ready times. Melouk et al. [32] developed a simulated annealing (SA) approach for minimizing $C_{\max}$ to schedule a batch processing machine with different job sizes [33]. Koh et al. proposed some heuristics and a random key-based representation GA approach for the problems of minimizing $C_{\max}$ and $\sum \omega_i C_i$ on a batch processing machine with incompatible job families and non-identical job sizes [34]. Kashan et al. proposed a hybrid genetic heuristic (HGH) to minimize makespan objective. To improve effectiveness of HGH, two local search heuristics are developed based on the problem characteristics for both batching and scheduling phases. The random batches procedure (RBP) is applied to ensure feasibility, and then for a feasible batching plan the longest processing time (LPT) rule is applied to schedule the batches on machines effectively [35]. Behnamian et al. proposed a hybrid heuristic for the problems with parallel machines and sequence-dependent setup times. The key innovative approach is the hybridization of an ACO, SA with VNS [36]. Wang et al. proposed a hybrid heuristic based on simulated annealing (SA) and genetic algorithm (GA) and a mixed integer programming (MIP) model [37]. Recently solving the batch machine scheduling problems using hybrid heuristic algorithms have been considered by researchers [38, 39].

4.3.2 Parallel Processing Machines Scheduling Problem

1. **Problem Description**

Suppose arbitrary number of jobs J from different job families F is available for processing on multiple batch machines at a given time. The jobs from the same family have the same processing time P_f. So the jobs from different families cannot be grouped into the same batch. This results in incompatibility of job families. All the batch machines M are identical and homogeneous in respect of processing times. The jobs cannot be interrupted during their processing. The problem is to schedule the jobs to parallel batch machines by satisfying the batch capacity constraints under the objective of minimizing $C_{\max}$.

The assumptions in the scheduling problem are as follows.

- All jobs sizes are arbitrary. All jobs have the same release dates or are available at a given time.

- All machines are homogeneous and identical in respect of job processing times. Each batch machine can process at most B jobs simultaneously. All batch machines are available at a given time.
- The jobs from the different job families cannot be processed together due to the different processing requirements for each job family.
- Once processing of a batch is started, it cannot be interrupted and no additional jobs can be introduced into the batch or removed from the batch until the processing is completed. So, only the jobs from the same family can be processed in a batch.

To obtain the optimal solutions, a mixed integer programming (MIP) model is presented. The nomenclature used in the process of developing MIP model is listed in Table 4.18.

The proposed MIP model is given as

$$\text{Min} \sum_{j=1}^{n} C_{\max}$$

Subject to:

$$\sum_{f=1}^{F} Y_{fbk} \leq 1$$

2. **Solution Approach**

To get the optimal or near optimal solutions, the batch machine scheduling problem is decomposed into two subproblems. As we known, if a complex problem is decomposed into many subproblems, there is more possibility of getting the solutions far away from global optimum, so each subproblem is solved without any further decomposition to void the more complex computation and get the global optimum in a reasonable time. The first subproblem is called as batch construction problem (BCP). The BCP is to assign the available jobs N_f from family f to the possible batches NB_f can be formed from the corresponding jobs. The second subproblem is called as batch scheduling problem (BSP). The BSP is to allocate the formed batches to the batch machines. Since the $C_{\max}$ of a particular machine depends on the sequence of the batches assigned to that machine, the sequences of the batches on each machine are to be considered. Let there are *NP* positions possible on each machine for the allocation of formed batches. So a total of *NB* batches are to be assigned to the possible *NP* positions on all *M* machines.

The BCP deals with construction of batches from the available jobs by satisfying the compatibility requirements of job families and batch size constraints. Since the jobs from different families cannot be grouped into the same batch, only the jobs from the same family will be batched. Generally, a batching plan with smaller summation of batch processing times PB_b in the batch constructing phase would result in a smaller makespan in batch scheduling phase. Hence, it can be inferred that job

Table 4.18 Meaning of symbols

Symbol	Meaning
F	Total number of families
f	Index for families, $f = 1, \ldots, F$
P_f	Processing time of each job from family f
N_f	Number of jobs in a family f
N	Total number of jobs, $N = \sum_{f=1}^{F} N_f$
J	Set of jobs, $J = \{1, \ldots, N\}$
j	Index for jobs, $\forall j \in J$
J_f	Set of jobs from family f, $J_f = \left\{\sum_{g=1}^{f-1} N_g + 1, \ldots, \sum_{g=1}^{f} N_g\right\}$
w_j	Weight or priorith of job j
M	Number of batch machines
I	Set of batch machines, $I = \{1, \ldots, M\}$
i	Index for batch machines, $\forall i \in I$
B	Capacity of batch machine
NB_f	Number of batches that can be formed from family f, $NB_f = N_f/C$
TB	Total number of batches that can formed, $TB = \sum_{f=1}^{F} NB_f$
SB	Set of batches that be formed, $SB = \{1, \ldots, TB\}$
b	Index for batches, $\forall b \in SB$
B_f	Set of batches that can be formed from family $f.B_f = \left\{\sum_{h=1}^{f-1} TB_h + 1, \ldots, \sum_{h=1}^{f} TB_h\right\}$
PB_b	Processing time of batch b, $PB_b = P_f, if\, b \in B_f, \forall f$
NP	Maximum number of positions possible for each machine, $NP = TB$
p	Index for position in the sequence of batches on each machine, $P = 1, \ldots, NP$
X_{jbi}	A judging variable, $X_{jbi} = \begin{cases} 1 & \text{if job } j \text{ is processed in batch } b \text{ of machine } i \\ 0 & \text{otherwise} \end{cases}$
Y_{fbi}	A judging variable, $Y_{fbi} = \begin{cases} 1 & \text{if family } f \text{ is processed in batch } b \text{ of machine } i \\ 0 & \text{otherwise} \end{cases}$

families with longer processing times in the same batch would be desirable. For the sake of longer processing time job families in the same batch, the heuristic named random batches rule (RBR) is proposed to construct a feasible batching plan. The RBR heuristic corresponded to the batching phase is as follows:

RBR Heuristic

BEGIN

Step I Assign all jobs to NB_f batches randomly without considering the capacity restriction;

Step II IF the derived batching plan is feasible THEN

Stop;

ELSE

Step 1 REPEAT

Choose a batch with capacity violation and the longest batch processing time. Select the job with the longest processing time in it;

Step 2 IF the selected job fits in any existing batch THEN

Step (1) IF there are some batches having the batch processing time longer than the processing time of the selected job THEN

Put the selected job in one of batches with the smallest residual capacity;

ELSE

Put the job in a feasible batch with the longest processing time;

END IF

ELSE

Create a new batch and assign it the selected job;

END IF

UNTIL getting a feasible batching plan

END IF

END

The BSP allocates formed batches to all machines. This means the allocation of each job to batch is known. This problem is equivalent to the parallel machine scheduling problems. The parallel machine scheduling problem is proved to be NP-hard. Hence, in this paper, an ACO algorithm is used to solve the BSP.

4.3.3 Ant Colony Optimization to Schedule Parallel Batch Processing Machines

1. **Solution construction**

For any combinatorial optimization, the solution to be constructed by an ant in ACO can be characterized by the solution components which construct the solution, the set of constraints to be followed by each solution components. The solution component for BSP will have two decisions to be made including machine selection first and then batch selection. Thus, in each iteration of ACO, an ant constructs the solution by building the solution components which consists of machine selection and batch selection. Each ant follows the constraints of the problem such as any batch should not be assigned to more than one machine. The cost associated with the present problem is the $C_{\max}$ which will be considered to update pheromone trail for the selection of batch. The details of the selection of batch machine and the pheromone trail update are described as follows.

2. **Batch machine selection**

In machine selection, an ant selects the machine by state transition rule as defined in ACO approach. Here, the greediness is the earliest availability of the machine. So the ant should select the earliest available machine which is incorporated as heuristic information for machine selection.

Let q be the random number generated from uniform distribution [0, 1]. The state transition rule of machine selection for an ant is defined as follows:

$$i = \begin{cases} \arg\max_{\forall m \in I}\{\lambda_m\}, & \text{if } q \le q_0 \\ I, & \text{otherwise} \end{cases} \tag{4.21}$$

In Eq. (4.21), the heuristic information is depicted in λ_m which is defined as the desirability of selecting a machine m. The heuristic information is the reciprocal of makespan (t_m) of machine m given by Eq. (4.22).

$$\lambda_m = \frac{1}{t_m}, \quad \forall m \in I \tag{4.22}$$

The probability q_0 is used in state transition rule to expand the search space of ants. An ant selects the machine with maximum desirability λ_i by favouring exploitation with probability q_0. With probability $(1-q_0)$ which favours the exploration the ant randomly selects the machine from the distribution formed by the probabilities P_i defined by Eq. (4.23).

$$P_i = \frac{\lambda_i}{\sum_{\forall m \in I} \lambda_m}, \quad \forall i \in I \tag{4.23}$$

3. **Batch selection**

Once the machine is selected, the next decision of an ant is to select the batch corresponding to the selected machine which defines a solution component for the BSP. The state transition rule for batch selection considers both the heuristic information and pheromone trails. The desirability between the selected machine and the formed batch to be allocated to that machine is the index as defined in RBR rule depicted as heuristic information. The selection of a batch for any machine depends not only on the desirability between the batch and the selected machine, but also on the attractiveness between the batch and the selected machine with respect to the whole solution. So pheromone trails are used in the batch selection. Assume that N is the set of unscheduled batches to the machine and r is the random number generated from uniform distribution [0, 1]. The state transition rule of batch selection for the selected machine i to be followed by ant is defined in Eq. (4.24).

$$b = \begin{cases} \arg\max_{\forall b \in N}\{\lambda_{ib}[\tau_{ib}]^{\alpha}\}, & \text{if } r \le r_0 \\ B, & \text{otherwise} \end{cases} \tag{4.24}$$

In Eq. (4.24), the heuristic information, λ_{ib}, is the desirability between the batch b and the selected machine i. The heuristic information is taken as the index. The pheromone trail, τ_{ib} introduced in Eq. (4.24) is described in detail in regard to how it is initialized, defined and updated in Sect. 5.4. The parameter α denotes the relative importance of pheromone trail as compared to heuristic information. λ_{ib} is defined as follows:

$$\lambda_{ib} = \pi_b \tag{4.25}$$

In Eq. (4.25), π_b is the index for batch b.

We introduce a biased exploration in batch selection with the probability $(1-r_0)$ in state transition rule as in the machine selection. Thus, the exploitation is favored with probability r_0 by selecting a batch b for the selected machine i with maximum desirability λ_{ib} To explore the decision of batch selection with the probability $(1-r_0)$, the ant randomly selects the batch from the probability distribution formed by the probabilities P_{ib} defined by Eq. (4.26).

$$P_{ib} = \begin{cases} \frac{\lambda_{ib}[\tau_{ib}]^{\alpha}}{\sum_{\forall k \in N} \lambda_{ik}[\tau_{ik}]^{\alpha}}, & \text{if } b \in N \\ 0, & \text{otherwise} \end{cases} \tag{4.26}$$

4. **Pheromone update**

The pheromone trail, τ_{ib} is defined as the desirability between the unscheduled batch $b \in \text{NB}$ and the selected machine I by the state transition rule in Eq. (4.24). Initially, there are no pheromone trails on solution components, so the pheromone trail on each solution component is set to a small value. Once an ant starts constructing a solution, the pheromone on the visited solution component is updated by 'local pheromone update' given in Eq. (4.27).

$$\tau_{ib} = (1 - \rho_{local})\tau_{ib} + \rho_{local}\tau_0 \tag{4.27}$$

Since the objective is the minimization of $C_{\max}$, τ_0 is defined as the reciprocal of RBR rule. ρ_{local} is the local pheromone evaporation rate ($0 < \rho_{local} < 1$) used for exploration by avoiding the unlimited accumulation of the pheromone trails and by forgetting the bad solutions. Thus, local pheromone update is applied to explore better solutions than the local optimal solutions.

Once a colony of ants has constructed feasible solutions, the global pheromone update is used to exploit the local best solution as given in Eq. (4.28). After each iteration of a colony of ants, the pheromone levels for the solution components which are in the best solution are updated with the value of $\Delta\tau_{ij}$ as in Eq. (4.29). As in the local pheromone update, ρ_{global} is the evaporation rate ($0 < \rho_{global} < 1$) used for exploration of the global best solution. But $\Delta\tau_{ij}$ favors the exploitation of the best local solution with a factor of Q on the reciprocal of the best objective value T^{best}.

$$\tau_{ib} = \left(1 - \rho_{global}\right)\tau_{ib} + \rho_{global}\tau_{ib} \tag{4.28}$$

where,

$$\Delta\tau_{ib} = \begin{cases} \frac{Q}{T^{best}}, & \text{if } ib \text{ is the best solution} \\ 0, & \text{otherwise} \end{cases} \tag{4.29}$$

5. **The overall algorithm**

In our algorithm, first the RBR rule is applied for BCP to form batches based on RBR index for jobs and then ACO algorithm is used for BSP to schedule the formed batches to the machines. The algorithm based on ACO can be summarized as follows.

Step I Batch construction (RBR rule)
Form batches using RBR rule for batch construction.
Step II Batch scheduling (ACO rule)
Step 1 Initialization
Set ACO parameters α, q_0, r_0, τ_0, β_{local}, ρ_{global}, Q, ITER-NUM (Total number of iterations), ANT-NUM (The number of ants in colony).
Initialize $\tau_{ib} = \tau_0\ \forall i \in I\ and\ b \in B.$
Set $p = 1$ (p = Iteration number).
Step 2 Solution construction
Set J and $l = 1$(The first ant in colony).
Step (1) Generate $q, r \sim U[0, 1]$.
Set $t_i = 0\ \forall\ i \in I$, where t_i is the completion time of the batches that have been already scheduled on machine $i \in$ I.
Step (2) Machine selection
Select the machine i by the following state transition rule,

$$i = \begin{cases} \arg\max_{\forall m \in I}\{\lambda_m\}, & \text{if } q \le q_0 \\ I, & \text{otherwise} \end{cases}$$

where, $\lambda_m = \frac{1}{t_m}$ and $I \sim$ Probability with $P_i = \lambda_i / \sum_{\forall m \in I} \lambda_m$
Step (3) Batch selection
Select the batch b for the selected machine i by the following state transition rule,

$$b = \begin{cases} \arg\max_{\forall b \in N}\{\lambda_{ib}[\tau_{ib}]^\alpha\}, & \text{if } r \le r_0 \\ B, & \text{otherwise} \end{cases}$$

where, $\lambda_{ib} = \pi_b$ and
$B \sim$ Probability with $P_{ib} = \frac{\lambda_{ib}[\tau_{ib}]^\alpha}{\sum_{\forall k \in N}\lambda_{ik}[\tau_{ik}]^\alpha}\forall b \in N.$
Update $N = N - b$.
Step (4) Local pheromone updation
Update the pheromone level for the selected machine i and batch b as:

$\tau_{ib} = (1 - \rho_{local})\tau_{ib} + \rho_{local}\tau_0$
If $N = \emptyset$, Go to Step (5). Otherwise, Go to Step (1).

Step (5) Evaluate current solution.
Find, T, the objective value of the current solution.
If $p = 1$, $T^{\text{best}} = \text{T}$.
If $p > 1$ and $T < T^{\text{best}}$, $T^{\text{best}} = T$.
Update $l = l + 1$ and $p = p + 1$.
If $l <$ ANT-NUM, Go to Step II. Otherwise, Go to Step (6).

Step (6) Global pheromone updation
Update the pheromone levels of the best solution found so far as:

$\tau_{ij} = \left(1 - \rho_{global}\right)\tau_{ij} + \rho_{global}\tau_{ib}$
where, $\Delta\tau_{ib} = Q/T^{best}$
Go to Step III.

Step 3 Termination criterion.
If $p \leq$ ITER-NUM, Go to Step II.
Otherwise, T^{best} is final solution and STOP.

1. Experiments and Results

For testing effectiveness of proposed algorithm, test problem instances are randomly generated in a manner similar to Chang et al. [28] For covering various types of problems, four factors are identified: number of jobs J, number of machines M, job sizes F, and job processing times P_f. Processing times and job sizes are generated from discrete uniform distribution. The factors, their levels and data, are shown in Table 4.19. The machine capacity B is assumed to be 10 in all instances.

Each category of problems is represented with a run code. For example a problem category with 50 jobs, two batch processing machines, processing times generated from the discrete uniform [1, 20] and job sizes generated from the discrete uniform [4, 8] is shown by J3m1p2f3.

The performance of an ACO algorithm depends on the initialization of ACO parameters, the problem data and size. It was observed that the choice of the parameters q_0, r_0, β_{local} and β_{global} is largely independent of the problem size and data. In the preliminary experimentation, 25 randomly generated problem instances were solved with the varied values of q_0, $r_0 \in \{0.8, 0.9, 0.99\}$, β_{local}, $\beta_{global} \in \{0.1, 0.01, 0.001\}$, $Q \in \{10{,}000, 50{,}000, 100{,}000\}$ and ANT-NUM $\in \{1, 5, 10, 20, 50\}$ and with the fixed values of $\alpha = 1$ and ITER-NUM $= 10{,}000$. The computational results for

Table 4.19 Factors and data

No	Factors	Data	Total values
1	J	10, 20 and 50	3
2	M	2, 4	2
3	P_f	Discrete uniform [1, 10] and discrete uniform [1, 20]	2
4	F	Discrete uniform [1, 10], discrete uniform [2, 4] and discrete uniform [4, 8]	3

this experiment indicated that the values of $q_0 = 0.9$, $r_0 = 0.9$, $\beta_{local} = 0.001$, $\beta_{global} = 0.001$, $Q = 10{,}000$ and ANT-NUM = 50 give good performance independent of the problem size. In the preliminary experimentation, it is observed that the quality of ACO approach.

For each problem size mainly depends on the selection of the parameters α and ITER-NUM. From the properties of ACO metaheuristic, the ACO performance will be almost stagnant after a certain number of iterations for a particular problem size and α value. Similarly for a particular problem size and ITER-NUM value, the ACO performance depends on the choice of α value. Now, we discuss the experimentation done to set the ACO parameters α and ITER-NUM. In the case of problems with the above test problem instances in Table 4.19, a total of 25 random problem instances are tested by varying the values of $\alpha\{1, 2, \ldots, 10\}$ and ITER-NUM $\in \{1000, 2000, 5000, 10{,}000, 20{,}000, \ldots, 100{,}000\}$. The results of the experimentation show that as ITER-NUM increases, the performance of the ACO increases over different α values. The performance of ACO increases significantly over $1000 \le$ ITER-NUM $\le 10{,}000$. But the performance increment is marginal for ITER-NUM > 10,000. It is also observed that for all the ITER-NUM values, $\alpha = 3$ gives good performance. So for the above test problem instances in Table 4.19, $\alpha = 3$ and ITER-NUM = 10,000.

The ACO and SA algorithm are coded in MATLAB 6.5.1 and all test problem instances are solved 20 times by ACO and SA on a Pentium IV, 2.2 GHz computer with 512 MB RAM.

Table 4.20 represents the results for test problem instances in Table 4.19. Columns 2, 3 and 4, report the best, average and the worst performance among the 15 runs of ACO, respectively. Column 5 reports the average time taken by ACO. These values are reported in columns 6, 7, 8 and 9 for SA, respectively. Column 10 compares ACO with SA in terms of solutions quality. The content of this column implies how much improvement ACO obtains over SA.

Table 4.20 Results for test problem instances

Run code	ACO				SA				Improvement (%)
	Best C_{max}	Avg C_{max}	Worst C_{max}	Avg. run time (s)	Best C_{max}	Avg C_{max}	Worst C_{max}	Avg. run time (s)	
$J_1m_1p_1f_1$	14	15.7	18	2.75	15	15.4	19	2.82	7.14
$J_1m_1p_1f_2$	11	13.3	16	1.45	11	13.1	18	1.34	0
$J_1m_1p_1f_3$	23	26	28	3.15	26	26	26	2.23	13.04
$J_1m_1p_2f_1$	19	20.8	21	1.65	21	21	21	1.47	10.53
$J_1m_1p_2f_2$	23	26.2	29	1.7	24	26	32	1.61	4.35
$J_1m_1p_2f_3$	49	52.1	55	2.05	52	52.1	54	1.85	6.12

(continued)

Table 4.20 (continued)

Run code	ACO				SA				Improvement (%)
	Best $C_{\max}$	Avg $C_{\max}$	Worst $C_{\max}$	Avg. run time (s)	Best $C_{\max}$	Avg $C_{\max}$	Worst $C_{\max}$	Avg. run time (s)	
$J_1m_2p_1f_1$	8	9.1	10	5.96	8	8.4	11	6.83	0
$J_1m_2p_1f_2$	9	9.8	11	4.42	9	9	9	4.57	0
$J_1m_2p_1f_3$	12	14.1	15	4.77	14	14	14	3.77	16.67
$J_1m_2p_2f_1$	17	17.9	19	2.15	19	19	19	1.65	11.76
$J_1m_2p_2f_2$	16	17.8	19	5.36	19	19	19	5.78	18.75
$J_1m_2p_2f_3$	48	49.8	52	2.1	52	52	52	1.8	8.33
$J_2m_1p_1f_1$	39	42.9	48	13.46	43	45.9	50	15.98	10.26
$J_2m_1p_1f_2$	19	24.5	32	7.83	24	26.5	31	8.49	26.32
$J_2m_1p_1f_3$	41	43.15	46	17.98	47	47.1	48	21.59	14.63
$J_2m_1p_2f_1$	59	72.3	82	18.97	72	75.9	82	19.67	22.03
$J_2m_1p_2f_2$	48	53.7	61	6.95	54	58.2	61	6.07	12.5
$J_2m_1p_2f_3$	92	95.8	109	16.73	104	104.8	106	18.73	13.04
$J_2m_2p_1f_1$	20	22.9	24	21.78	22	23.3	25	28.38	10
$J_2m_2p_1f_2$	13	14.2	15	18.86	13	13.4	15	18.86	0
$J_2m_2p_1f_3$	21	24.3	27	26.69	24	24.3	26	28.09	14.29
$J_2m_2p_2f_1$	32	34	41	48.71	36	39	45	51.74	12.5
$J_2m_2p_2f_2$	25	25.2	28	12.5	28	30.2	31	13.9	12
$J_2m_2p_2f_3$	48	54.5	56	25.76	54	54.5	56	26.16	12.5
$J_3m_1p_1f_1$	54	62.7	715	148.53	67	68.7	75	169.39	24.07
$J_3m_1p_1f_2$	42	53.5	61	53.02	50	53.5	56	55.78	19.05
$J_3m_1p_1f_3$	85	93.2	97	123.5	92	95.2	99	130.5	8.24
$J_3m_1p_2f_1$	129	139.9	169	139.8	156	161.9	177	148.6	20.94
$J_3m_1p_2f_2$	87	94.1	104	42.97	96	100.1	104	47.7	10.34
$J_3m_1p_2f_3$	159	156.9	161	99.8	181	182.9	185	112.6	13.84
$J_3m_2p_1f_1$	28	34.4	37	378.46	34	34.4	36	390.53	21.43
$J_3m_2p_1f_2$	23	26.2	28	111.9	25	26.9	29	120.9	8.7
$J_3m_2p_1f_3$	42	47.8	49	229.7	47	47.8	50	240.3	11.9
$J_3m_2p_2f_1$	72	78.9	854	377.3	77	80.9	85	377.3	6.94
$J_3m_2p_2f_2$	44	50.1	54	118.1	49	50.9	53	118.1	11.36
$J_3m_2p_2f_3$	89	92.1	94	130	91	92	93	130	2.25

Appendix 1

The Whole Calculation of GMT Method

	W_1	W_2	W_3	W_4	W_5	W_6	W_1	W_2	W_3	W_4	W_5	W_6	W_1	W_2	W_3	W_4	W_5	W_6
O_{11}	2	3	X	5	**2**	3	2	3	X	5	**2**	3	**4**	3	X	5	2	3
O_{12}	4	3	5	X	3	X	4	3	5	X	**5**	X	**6**	3	5	X	5	X
O_{13}	2	X	5	4	X	4	**2**	X	5	4	**X**	4	**2**	X	5	4	X	4
O_{21}	3	X	5	3	2	X	3	X	5	3	**4**	X	**5**	X	5	3	4	X
O_{22}	4	X	3	3	X	5	4	X	3	3	**X**	5	**6**	X	3	3	X	5
O_{23}	X	X	4	5	7	9	X	X	4	5	**9**	9	**X**	X	4	5	9	9
O_{31}	5	6	X	4	X	4	5	6	X	4	**X**	4	**7**	6	X	4	X	4
O_{32}	X	4	X	3	5	4	X	4	X	3	**7**	4	**X**	4	X	3	7	4
O_{33}	X	X	11	X	9	13	X	X	11	X	**11**	13	**X**	X	11	X	11	13
O_{41}	9	X	7	9	X	6	9	X	7	9	**X**	6	**11**	X	7	9	X	6
O_{42}	X	7	X	5	6	5	X	7	X	5	**8**	5	**X**	7	X	5	8	5
O_{43}	2	3	4	X	X	4	2	3	4	X	**X**	4	**4**	3	4	X	X	4
O_{51}	X	4	5	3	X	4	X	4	5	3	**X**	4	**X**	4	5	3	X	4
O_{52}	4	4	6	X	3	5	4	4	6	X	**5**	5	**6**	4	6	X	5	5
O_{53}	3	4	X	5	6	X	3	4	X	5	**8**	X	**5**	4	X	5	8	X
O_{61}	X	3	7	4	5	X	X	3	7	4	**7**	X	**X**	3	7	4	7	X
O_{62}	6	2	X	4	3	X	6	2	X	4	**5**	X	**8**	**2**	X	4	5	X
O_{63}	5	4	3	X	X	4	5	4	3	X	**X**	4	**7**	4	3	X	X	4
	W_1	W_2	W_3	W_4	W_5	W_6	W_1	W_2	W_3	W_4	W_5	W_6	W_1	W_2	W_3	W_4	W_5	W_6
O_{11}	4	**5**	X	5	2	3	4	5	X	**8**	2	3	4	5	**X**	8	2	3
O_{12}	6	**5**	5	X	5	X	6	5	5	**X**	5	X	6	5	**8**	X	5	X
O_{13}	2	**X**	5	4	X	4	2	X	5	**7**	X	4	2	X	**8**	7	X	4
O_{21}	5	**X**	5	**3**	4	X	5	X	5	**3**	4	X	5	X	**8**	3	4	X
O_{22}	6	**X**	3	3	X	5	6	X	3	**6**	X	5	6	X	**6**	6	X	5
O_{23}	X	**X**	4	5	9	9	X	X	4	**8**	9	9	X	X	**7**	8	9	9
O_{31}	7	**8**	X	4	X	4	7	8	X	**7**	X	4	7	8	**X**	7	X	**4**
O_{32}	X	**6**	X	3	7	4	X	6	X	**6**	7	4	X	6	**X**	6	7	4
O_{33}	X	**X**	11	X	11	13	X	X	11	**X**	11	13	X	X	**14**	X	11	13
O_{41}	11	**X**	7	9	X	6	11	X	7	**12**	X	6	11	X	**10**	12	X	6
O_{42}	X	**9**	X	5	8	5	X	9	X	**8**	8	5	X	9	**X**	8	8	5
O_{43}	4	**5**	4	X	X	4	4	5	4	**X**	X	4	4	5	**7**	X	X	4
O_{51}	X	**6**	5	3	X	4	X	6	5	**6**	X	4	X	6	**8**	6	X	4
O_{52}	6	**6**	6	X	5	5	6	6	6	**X**	5	5	6	6	**9**	X	5	5
O_{53}	5	**6**	X	5	8	X	5	6	X	**8**	8	X	5	6	**X**	8	8	X

(continued)

(continued)

	W_1	W_2	W_3	W_4	W_5	W_6	W_1	W_2	W_3	W_4	W_5	W_6	W_1	W_2	W_3	W_4	W_5	W_6
O_{61}	X	**5**	7	4	7	X	X	5	7	**7**	7	X	X	5	**10**	7	7	X
O_{62}	8	**2**	X	4	5	X	8	2	X	**7**	5	X	8	2	**X**	7	5	X
O_{63}	7	**6**	3	X	X	4	7	6	**3**	**X**	X	4	7	6	**3**	X	X	4
	W_1	W_2	W_3	W_4	W_5	W_6	W_1	W_2	W_3	W_4	W_5	W_6	W_1	W_2	W_3	W_4	W_5	W_6
O_{11}	4	5	X	8	2	**7**	**8**	5	X	8	2	7	8	**10**	X	8	2	7
O_{12}	6	5	8	X	5	**X**	**10**	**5**	8	X	5	X	10	**5**	8	X	5	X
O_{13}	2	X	8	7	X	**8**	**6**	X	8	7	X	8	6	**X**	8	7	X	8
O_{21}	5	X	8	3	4	**X**	**9**	X	8	3	4	X	9	**X**	8	3	4	X
O_{22}	6	X	6	6	X	**9**	**10**	X	6	6	X	9	10	**X**	6	6	X	9
O_{23}	X	X	7	8	9	**13**	**X**	X	7	8	9	13	X	**X**	7	8	9	13
O_{31}	7	8	X	7	X	**4**	**11**	8	X	7	X	4	11	**13**	X	7	X	4
O_{32}	X	6	X	6	7	**8**	**X**	6	X	6	7	8	X	**11**	X	6	7	8
O_{33}	X	X	14	X	11	**17**	**X**	X	14	X	11	17	X	**X**	14	X	11	17
O_{41}	11	X	10	12	X	**10**	**15**	X	10	12	X	10	15	**X**	10	12	X	10
O_{42}	X	9	X	8	8	**9**	**X**	9	X	8	8	9	X	**14**	X	8	8	9
O_{43}	**4**	5	7	X	X	**8**	**4**	5	7	X	X	8	4	**10**	7	X	X	8
O_{51}	X	6	8	6	X	**8**	**X**	6	8	6	X	8	X	**11**	8	6	X	8
O_{52}	6	6	9	X	5	**9**	**10**	6	9	X	5	9	10	**11**	9	X	**5**	9
O_{53}	5	6	X	8	8	**X**	**9**	6	X	8	8	X	9	**11**	X	8	8	X
O_{61}	X	5	10	7	7	**X**	**X**	5	10	7	7	X	X	**10**	10	7	7	X
O_{62}	8	2	X	7	5	**X**	**12**	2	X	7	5	X	12	**7**	X	7	5	X
O_{63}	7	6	3	X	X	**8**	**11**	6	3	X	X	8	11	**11**	3	X	X	8
	W_1	W_2	W_3	W_4	W_5	W_6	W_1	W_2	W_3	W_4	W_5	W_6	W_1	W_2	W_3	W_4	W_5	W_6
O_{11}	8	10	X	8	**7**	7	8	10	**X**	8	7	7	8	10	X	**14**	7	7
O_{12}	10	5	8	X	**10**	X	10	5	**14**	X	10	X	10	5	14	**X**	10	X
O_{13}	6	X	8	7	**X**	8	6	X	**14**	7	X	8	6	X	14	**13**	X	8
O_{21}	9	X	8	3	**9**	X	9	X	**14**	3	9	X	9	X	14	**9**	9	X
O_{22}	10	X	**6**	6	**X**	9	10	X	**6**	6	X	9	10	X	6	**12**	X	9
O_{23}	X	X	7	8	**14**	13	X	X	**13**	8	14	13	X	X	13	**14**	14	13
O_{31}	11	13	X	7	**X**	4	11	13	**X**	7	X	4	11	13	X	**13**	X	4
O_{32}	X	11	X	6	**12**	8	X	11	**X**	**6**	12	8	X	11	X	**6**	12	8
O_{33}	X	X	14	X	**16**	17	X	X	**20**	X	16	17	X	X	20	**X**	16	17
O_{41}	15	X	10	12	**X**	10	15	X	**16**	12	X	10	15	X	16	**18**	X	10
O_{42}	X	14	X	8	**13**	9	X	14	**X**	8	13	9	X	14	X	**14**	13	9
O_{43}	4	10	7	X	**X**	8	4	10	**13**	X	X	8	4	10	13	**X**	X	8
O_{51}	X	11	8	6	**X**	8	X	11	**14**	6	X	8	X	11	14	**12**	X	**8**
O_{52}	10	11	9	X	**5**	9	10	11	**15**	X	10	9	10	11	15	**X**	10	9
O_{53}	9	11	X	8	**13**	X	9	11	**X**	8	13	X	9	11	X	**14**	13	X

(continued)

(continued)

	W_1	W_2	W_3	W_4	W_5	W_6	W_1	W_2	W_3	W_4	W_5	W_6	W_1	W_2	W_3	W_4	W_5	W_6
O_{61}	X	10	10	7	**12**	X	X	10	**16**	7	12	X	X	10	16	**13**	12	X
O_{62}	12	7	X	7	**10**	X	12	7	**X**	7	10	X	12	7	X	**13**	10	X
O_{63}	11	11	3	X	**X**	8	11	11	**9**	X	X	8	11	11	9	**X**	X	8
	W_1	W_2	W_3	W_4	W_5	W_6	W_1	W_2	W_3	W_4	W_5	W_6	W_1	W_2	W_3	W_4	W_5	W_6
O_{11}	8	10	X	14	7	**15**	**17**	10	X	14	7	15	17	**20**	X	14	7	15
O_{12}	10	5	14	X	10	**X**	**19**	5	14	X	10	X	19	**15**	14	X	10	X
O_{13}	6	X	14	13	X	**16**	**15**	X	14	13	X	16	15	**X**	14	13	X	16
O_{21}	9	X	14	9	9	**X**	**18**	X	14	9	9	X	18	**X**	14	9	9	X
O_{22}	10	X	6	12	X	**17**	**19**	X	6	12	X	17	19	**X**	6	12	X	17
O_{23}	X	X	13	14	14	**21**	**X**	X	13	14	14	21	X	**X**	**13**	14	14	21
O_{31}	11	13	X	13	X	**12**	**20**	13	X	13	X	12	20	**23**	X	13	X	12
O_{32}	X	11	X	6	12	**16**	**X**	11	X	6	12	16	X	**21**	X	6	12	16
O_{33}	X	X	20	X	16	**25**	**X**	X	20	X	16	25	X	**X**	20	X	16	25
O_{41}	15	X	16	18	X	**18**	**24**	X	16	18	X	18	24	**X**	16	18	X	18
O_{42}	X	14	X	14	13	**17**	**X**	14	X	14	13	17	X	**24**	X	14	13	17
O_{43}	4	10	13	X	X	**16**	**13**	10	13	X	X	16	13	**20**	13	X	X	16
O_{51}	X	11	14	12	X	**8**	**X**	11	14	12	X	8	X	**21**	14	12	X	8
O_{52}	10	11	15	X	10	**17**	**19**	11	15	X	10	17	19	**21**	15	X	10	17
O_{53}	**9**	11	X	14	13	**X**	**9**	11	X	14	13	X	9	**21**	X	14	13	X
O_{61}	X	10	16	13	12	**X**	**X**	**10**	16	13	12	X	X	**10**	16	13	12	X
O_{62}	12	7	X	13	10	**X**	**21**	7	X	13	10	X	21	**17**	X	13	10	X
O_{63}	11	11	9	X	X	**16**	**20**	11	9	X	X	16	20	**21**	9	X	X	16
	W_1	W_2	W_3	W_4	W_5	W_6	W_1	W_2	W_3	W_4	W_5	W_6	W_1	W_2	W_3	W_4	W_5	W_6
O_{11}	17	20	**X**	14	7	15	17	20	X	14	**20**	15	17	20	X	**32**	20	15
O_{12}	19	15	**27**	X	10	X	19	15	27	X	**23**	X	19	15	27	**X**	23	X
O_{13}	15	X	**27**	13	X	16	15	X	27	13	**X**	16	15	X	27	**31**	X	16
O_{21}	18	X	**27**	9	9	X	18	X	27	9	**22**	X	18	X	27	**27**	22	X
O_{22}	19	X	**19**	12	X	17	19	X	19	12	**X**	17	19	X	19	**30**	X	17
O_{23}	X	X	**13**	14	14	21	X	X	13	14	**27**	21	X	X	13	**32**	27	21
O_{31}	20	23	**X**	13	X	12	20	23	X	13	**X**	12	20	23	X	**31**	X	12
O_{32}	X	21	**X**	6	12	16	X	21	X	6	**25**	16	X	21	X	**24**	25	16
O_{33}	X	X	**33**	X	16	25	X	X	33	X	**29**	25	X	X	33	**X**	29	25
O_{41}	24	X	**29**	18	X	18	24	X	29	**18**	**X**	18	24	X	29	**18**	X	18
O_{42}	X	24	**X**	14	**13**	17	X	24	X	14	**13**	17	X	24	X	**32**	13	17
O_{43}	13	20	**26**	X	X	16	13	20	26	X	**X**	16	13	20	26	**X**	X	16
O_{51}	X	21	**27**	12	X	8	X	21	27	12	**X**	8	X	21	27	**30**	X	8
O_{52}	19	21	**28**	X	10	17	19	21	28	X	**23**	17	19	21	28	**X**	23	17
O_{53}	9	21	**X**	14	13	X	9	21	X	14	**26**	X	9	21	X	**32**	26	X
O_{61}	X	10	**29**	13	12	X	X	10	29	13	**25**	X	X	10	29	**31**	25	X

(continued)

(continued)

	W_1	W_2	W_3	W_4	W_5	W_6	W_1	W_2	W_3	W_4	W_5	W_6	W_1	W_2	W_3	W_4	W_5	W_6
O_{62}	21	17	**X**	13	10	X	21	17	X	13	**23**	X	21	17	X	**31**	23	X
O_{63}	20	21	**22**	X	X	16	20	21	22	X	**X**	16	20	21	22	**X**	X	16

Appendix 2

The Whole Calculation of RS Method

	W_1	W_2	W_3	W_4	W_5	W_6	W_1	W_2	W_3	W_4	W_5	W_6	W_1	W_2	W_3	W_4	W_5	W_6
O_{11}	2	3	X	5	**2**	3	2	3	X	5	**2**	3	2	3	X	5	2	3
O_{21}	3	X	5	3	2	X	3	X	5	**3**	**4**	X	3	X	5	**3**	4	X
O_{31}	5	6	X	4	X	4	5	6	X	4	**X**	4	5	6	X	**7**	X	**4**
O_{41}	9	X	7	9	X	6	9	X	7	9	**X**	6	9	X	7	**12**	X	6
O_{51}	X	4	5	3	X	4	X	4	5	3	**X**	4	X	4	5	**6**	X	4
O_{61}	X	3	7	4	5	X	X	3	7	4	**7**	X	X	3	7	**7**	7	X
O_{42}	X	7	X	5	6	5	X	7	X	5	**8**	5	X	7	X	**8**	8	5
O_{12}	4	3	5	X	3	X	4	3	5	X	**5**	X	4	3	5	**X**	5	X
O_{52}	4	4	6	X	3	5	4	4	6	X	**5**	5	4	4	6	**X**	5	5
O_{22}	4	X	3	3	X	5	4	X	3	3	**X**	5	4	X	3	**6**	X	5
O_{13}	2	X	5	4	X	4	2	X	5	4	**X**	4	2	X	5	**7**	X	4
O_{43}	2	3	4	X	X	4	2	3	4	X	**X**	4	2	3	4	**X**	X	4
O_{62}	6	2	X	4	3	X	6	2	X	4	**5**	X	6	2	X	**7**	5	X
O_{53}	3	4	X	5	6	X	3	4	X	5	**8**	X	3	4	X	**8**	8	X
O_{23}	X	X	4	5	7	9	X	X	4	5	**9**	9	X	X	4	**8**	9	9
O_{63}	5	4	3	X	X	4	5	4	3	X	**X**	4	5	4	3	**X**	X	4
O_{32}	X	4	X	3	5	4	X	4	X	3	**7**	4	X	4	X	**6**	7	4
O_{33}	X	X	11	X	9	13	X	X	11	X	**11**	13	X	X	11	**X**	11	13
	W_1	W_2	W_3	W_4	W_5	W_6	W_1	W_2	W_3	W_4	W_5	W_6	W_1	W_2	W_3	W_4	W_5	W_6
O_{11}	2	3	X	5	2	3	2	3	X	5	2	3	2	3	X	5	2	3
O_{21}	3	X	5	3	4	X	3	X	5	3	4	X	3	X	5	3	4	X
O_{31}	5	6	X	7	X	**4**	5	6	X	7	X	4	5	6	X	7	X	4
O_{41}	9	X	**7**	12	X	**10**	9	X	**7**	12	X	10	9	X	7	12	X	10
O_{51}	X	4	5	6	X	**8**	X	**4**	**12**	6	X	8	X	**4**	12	6	X	8
O_{61}	X	3	7	7	7	**X**	X	3	**14**	7	7	X	X	**7**	14	7	**7**	X
O_{42}	X	7	X	8	8	**9**	X	7	**X**	8	8	9	X	**11**	X	8	8	9
O_{12}	4	3	5	X	5	**X**	4	3	**12**	X	5	X	4	**7**	12	X	5	X
O_{52}	4	4	6	X	5	**9**	4	4	**13**	X	5	9	4	**8**	13	X	5	9

(continued)

(continued)

	W_1	W_2	W_3	W_4	W_5	W_6	W_1	W_2	W_3	W_4	W_5	W_6	W_1	W_2	W_3	W_4	W_5	W_6
O_{22}	4	X	3	6	X	**9**	4	X	**10**	6	X	9	4	**X**	10	6	X	9
O_{13}	2	X	5	7	X	**8**	2	X	**12**	7	X	8	2	**X**	12	7	X	8
O_{43}	2	3	4	X	X	**8**	2	3	**11**	X	X	8	2	**7**	11	X	X	8
O_{62}	6	2	X	7	5	**X**	6	2	**X**	7	5	X	6	**6**	X	7	5	X
O_{53}	3	4	X	8	8	**X**	3	4	**X**	8	8	X	3	**8**	X	8	8	X
O_{23}	X	X	4	8	9	**13**	X	X	**11**	8	9	13	X	**X**	11	8	9	13
O_{63}	5	4	3	X	X	**8**	5	4	**10**	X	X	8	5	**8**	10	X	X	8
O_{32}	X	4	X	6	7	**8**	X	4	**X**	6	7	8	X	**8**	X	6	7	8
O_{33}	X	X	11	X	11	**17**	X	X	**18**	X	11	17	X	**X**	18	X	11	17
	W_1	W_2	W_3	W_4	W_5	W_6	W_1	W_2	W_3	W_4	W_5	W_6	W_1	W_2	W_3	W_4	W_5	W_6
O_{11}	2	3	X	5	2	3	2	3	X	5	2	3	2	3	X	5	2	3
O_{21}	3	X	5	3	4	X	3	X	5	3	4	X	3	X	5	3	4	X
O_{31}	5	6	X	7	X	4	5	6	X	7	X	4	5	6	X	7	X	4
O_{41}	9	X	7	12	X	10	9	X	7	12	X	10	9	X	7	12	X	10
O_{51}	X	4	12	6	X	8	X	4	12	6	X	8	X	4	12	6	X	8
O_{61}	X	7	14	7	**7**	X	X	7	14	7	7	X	X	7	14	7	7	X
O_{42}	X	11	X	**8**	**15**	9	X	11	X	**8**	15	9	X	11	X	8	15	9
O_{12}	4	7	12	X	**12**	X	**4**	7	12	**X**	12	X	**4**	7	12	X	12	X
O_{52}	4	8	13	X	**12**	9	4	8	13	**X**	12	9	**8**	8	13	X	12	9
O_{22}	4	X	10	6	**X**	9	4	X	10	**14**	X	9	**8**	X	10	14	X	9
O_{13}	2	X	12	7	**X**	8	2	X	12	**15**	X	8	**6**	X	12	15	X	8
O_{43}	2	7	11	X	**X**	8	2	7	11	**X**	X	8	**6**	7	11	X	X	8
O_{62}	6	6	X	7	**12**	X	6	6	X	**15**	12	X	**10**	6	X	15	12	X
O_{53}	3	8	X	8	**15**	X	3	8	X	**16**	15	X	**7**	8	X	16	15	X
O_{23}	X	X	11	8	**16**	13	X	X	11	**16**	16	13	**X**	X	11	16	16	13
O_{63}	5	8	10	X	**X**	8	5	8	10	**X**	X	8	**9**	8	10	X	X	8
O_{32}	X	8	X	6	**14**	8	X	8	X	**14**	14	8	**X**	8	X	14	14	8
O_{33}	X	X	18	X	**18**	17	X	X	18	**X**	18	17	**X**	X	18	X	18	17
	W_1	W_2	W_3	W_4	W_5	W_6	W_1	W_2	W_3	W_4	W_5	W_6	W_1	W_2	W_3	W_4	W_5	W_6
O_{11}	2	3	X	5	2	3	2	3	X	5	2	3	2	3	X	5	2	3
O_{21}	3	X	5	3	4	X	3	X	5	3	4	X	3	X	5	3	4	X
O_{31}	5	6	X	7	X	4	5	6	X	7	X	4	5	6	X	7	X	4
O_{41}	9	X	7	12	X	10	9	X	7	12	X	10	9	X	7	12	X	10
O_{51}	X	4	12	6	X	8	X	4	12	6	X	8	X	4	5	6	X	8
O_{61}	X	7	14	7	7	X	X	7	14	7	7	X	X	7	7	7	7	X
O_{42}	X	11	X	8	15	9	X	11	X	8	15	9	X	11	X	8	15	9
O_{12}	4	7	12	X	12	X	4	7	12	X	12	X	4	7	5	X	12	X
O_{52}	**8**	8	13	X	12	9	8	8	13	X	12	9	8	8	6	X	12	9
O_{22}	**16**	X	10	14	X	**9**	16	X	10	14	X	**9**	8	X	9	14	X	9

(continued)

(continued)

	W_1	W_2	W_3	W_4	W_5	W_6	W_1	W_2	W_3	W_4	W_5	W_6	W_1	W_2	W_3	W_4	W_5	W_6
O_{13}	**14**	X	12	15	X	8	14	X	**12**	15	X	**17**	14	X	**12**	15	X	17
O_{43}	**14**	7	11	X	X	8	14	7	11	X	X	**17**	14	**7**	**23**	X	X	17
O_{62}	**18**	6	X	15	12	X	18	6	X	15	12	**X**	18	6	**X**	15	12	X
O_{53}	**15**	8	X	16	15	X	15	8	X	16	15	**X**	15	8	**X**	16	15	X
O_{23}	**X**	X	11	16	16	13	X	X	11	16	16	**22**	X	X	**23**	16	16	22
O_{63}	**17**	8	10	X	X	8	17	8	10	X	X	**17**	17	8	**22**	X	X	17
O_{32}	**X**	8	X	14	14	8	X	8	X	14	14	**17**	X	8	**X**	14	14	17
O_{33}	**X**	X	18	X	18	17	X	X	18	X	18	**26**	X	X	**30**	X	18	26
	W_1	W_2	W_3	W_4	W_5	W_6	W_1	W_2	W_3	W_4	W_5	W_6	W_1	W_2	W_3	W_4	W_5	W_6
O_{11}	2	3	X	5	2	3	2	3	X	5	2	3	2	3	X	5	2	3
O_{21}	3	X	5	3	4	X	3	X	5	3	4	X	3	X	5	3	4	X
O_{31}	5	6	X	7	X	4	5	6	X	7	X	4	5	6	X	7	X	4
O_{41}	9	X	7	12	X	10	9	X	7	12	X	10	9	X	7	12	X	10
O_{51}	X	4	5	6	X	8	X	4	5	6	X	8	X	4	5	6	X	8
O_{61}	X	7	7	7	7	X	X	7	7	7	7	X	X	7	7	7	7	X
O_{42}	X	11	X	8	15	9	X	11	X	8	15	9	X	11	X	8	15	9
O_{12}	4	7	5	X	12	X	4	7	5	X	12	X	4	7	5	X	12	X
O_{52}	8	8	6	X	12	9	8	8	6	X	12	9	8	8	6	X	12	9
O_{22}	8	X	9	14	X	9	8	X	9	14	X	9	8	X	9	14	X	9
O_{13}	14	X	12	15	X	17	14	X	12	15	X	17	14	X	12	15	X	17
O_{43}	14	**7**	23	X	X	17	14	7	23	X	X	17	14	7	23	X	X	17
O_{62}	18	**13**	X	15	**12**	X	18	13	X	15	**12**	X	18	13	X	15	12	X
O_{53}	15	**15**	X	16	15	X	15	**15**	X	16	**27**	X	15	**15**	X	16	27	X
O_{23}	X	**X**	23	16	16	22	X	X	23	16	**28**	22	X	**X**	23	**16**	28	22
O_{63}	17	**15**	22	X	X	17	17	15	22	X	**X**	17	17	**30**	22	X	X	17
O_{32}	X	**15**	X	14	14	17	X	15	X	14	**26**	17	X	**30**	X	14	26	17
O_{33}	X	**X**	30	X	18	26	X	X	30	X	**30**	26	X	**X**	30	X	30	26
	W_1	W_2	W_3	W_4	W_5	W_6	W_1	W_2	W_3	W_4	W_5	W_6	W_1	W_2	W_3	W_4	W_5	W_6
O_{11}	2	3	X	5	2	3	2	3	X	5	2	X	2	3	X	5	2	3
O_{21}	3	X	5	3	4	X	3	X	5	3	4	X	3	X	5	3	4	X
O_{31}	5	6	X	7	X	4	5	6	X	7	X	4	5	6	X	7	X	4
O_{41}	9	X	7	12	X	10	9	X	7	12	X	10	9	X	7	12	X	10
O_{51}	X	4	5	6	X	8	X	4	5	6	X	8	X	4	5	6	X	8
O_{61}	X	7	7	7	7	X	X	7	7	7	7	X	X	7	7	7	7	X
O_{42}	X	11	X	8	15	9	X	11	X	8	15	9	X	11	X	8	15	9
O_{12}	4	7	5	X	12	X	4	7	5	X	12	X	4	7	5	X	12	X
O_{52}	8	8	6	X	12	9	8	8	6	X	12	9	8	8	6	X	12	9
O_{22}	8	X	9	14	X	9	8	X	9	14	X	9	8	X	9	14	X	9
O_{13}	14	X	12	15	X	17	14	X	11	15	X	8	14	X	11	15	X	8

(continued)

(continued)

	W_1	W_2	W_3	W_4	W_5	W_6	W_1	W_2	W_3	W_4	W_5	W_6	W_1	W_2	W_3	W_4	W_5	W_6
O_{43}	14	7	23	X	X	17	14	7	10	X	X	16	14	7	10	X	X	16
O_{62}	18	13	X	15	12	X	18	13	X	15	12	X	18	13	X	15	12	X
O_{53}	15	15	X	16	27	X	15	15	X	16	27	X	15	15	X	16	27	X
O_{23}	X	X	23	**16**	28	22	X	X	23	16	28	21	X	X	21	16	28	21
O_{63}	**17**	30	22	$\boldsymbol{X}$	X	17	**17**	30	22	X	X	17	17	30	20	X	X	17
O_{32}	X	30	X	**30**	26	**17**	$\boldsymbol{X}$	30	X	30	26	17	X	30	X	30	26	**17**
O_{33}	X	X	30	$\boldsymbol{X}$	30	26	$\boldsymbol{X}$	X	30	X	30	26	X	X	30	X	30	**41**

References

1. Xia W, Zhiming W (2005) An effective hybrid optimization approach for multi-objective flexible job-shop scheduling problems. Comput Ind Eng 48(2):409–425
2. Broker P, Schlie R (1990) Job-shop scheduling with multi-purpose machines. Computing 45(4):369–375
3. Brandimarte P (1993) Routing and scheduling in a flexible job shop by taboo search. Ann Oper Res 41:157–183
4. Paulli J (1995) A hierarchical approach for the FMS scheduling problem. Eur J Oper Res 86(1):32–42
5. Shi-Jin W, Bing-Hai Z, Li-Feng X (2008) A filtered-beam-search-based heuristic algorithm for flexible job-shop scheduling problem. Int J Prod Res 46(11):3027–3058
6. Kacem I, Hammadi S, Borne P (2002) Approach by localization and multiobjective evolutionary optimization for flexible job-shop scheduling problems. IEEE Trans Syst Man Cybern Part C 3(21):1–13
7. Kacem I, Hammadi S, Borne P (2002) Pareto-optimality approach for flexible job-shop scheduling problems: hybridization of evolutionary algorithms and fuzzy logic. Math Comput Simul 60(3–5):245–276
8. Lei D (2010) A genetic algorithm for flexible job shop scheduling with fuzzy processing time. Int J Prod Res 48(10):2995–3013
9. Hurink E, Jurisch B, Thole M (1994) Tabu search for the job shop scheduling problem with multi-purpose machines. Oper Res Spektrum 15:205–215
10. Dauzere-Peres S, Paulli J (1997) An integrated approach for modeling and solving the general multiprocessor job-shop scheduling problem using tabu search. Ann Oper Res 70:281–306
11. Mastrolilli M, Gambardella LM (2002) Effective neighborhood functions for the flexible job shop problem. J Sched 3(1):3–20
12. Amiri M, Zandieh M, Yazdani M, Bagheri A (2010) A variable neighbourhood search algorithm for the flexible job-shop scheduling problem. Int J Prod Res 48(19):5671–5689
13. Lu PH, Wu MC, Tan H et al (2018) A genetic algorithm embedded with a concise chromosome representation for distributed and flexible job-shop scheduling problems. J Intell Manuf 29:19–34. https://doi.org/10.1007/s10845-015-1083-z
14. Chang HC, Liu TK (2017) Optimisation of distributed manufacturing flexible job shop scheduling by using hybrid genetic algorithms. J Intell Manuf 28:1973–1986. https://doi.org/10.1007/s10845-015-1084-y
15. Xu W, Hu Y, Luo W, Wang L, Wu R (2021) A multi-objective scheduling method for distributed and flexible job shop based on hybrid genetic algorithm and tabu search considering operation outsourcing and carbon emission. Comput Ind Eng 157:107318. https://doi.org/10.1016/j.cie.2021.107318

16. Krzysztof K, Krenczyk D, Paprocka I, Kempa WM, Grabowik C (2017) Ant colony optimisation for scheduling of flexible job shop with multi-resources requirements. MATEC Web Conf 112:06018. https://doi.org/10.1051/matecconf/201711206018
17. Nouiri M, Bekrar A, Jemai A et al (2018) An effective and distributed particle swarm optimization algorithm for flexible job-shop scheduling problem. J Intell Manuf 29:603–615. https://doi.org/10.1007/s10845-015-1039-3
18. Meng T, Pan Q-K, Sang H-Y (2018) A hybrid artificial bee colony algorithm for a flexible job shop scheduling problem with overlapping in operations. Int J Prod Res 56(16):5278–5292. https://doi.org/10.1080/00207543.2018.1467575
19. Li X, Xie J, Ma Q et al (2022) Improved gray wolf optimizer for distributed flexible job shop scheduling problem. Sci China Technol Sci 65:2105–2115. https://doi.org/10.1007/s11431-022-2096-6
20. Gaham M, Bouzouia B, Achour N (2018) An effective operations permutation-based discrete harmony search approach for the flexible job shop scheduling problem with makespan criterion. Appl Intell 48:1423–1441. https://doi.org/10.1007/s10489-017-0993-1
21. Dobson G, Nambimadom RS (2001) The batch loading and scheduling problem. Oper Res 49:52–65
22. Webster S, Baker KR (1995) Scheduling groups of jobs on a single machine. Oper Res 43(4):692–703
23. Dileepan P, Ahmadi M (2010) Scheduling rules for a small dynamic job-shop: a simulation approach. Int J Simul Model 9(4):173–183
24. Lee C-Y, Uzsoy R, Martin-Vega L-A (1992) Efficient algorithms for scheduling semiconductor burn-in operations. Oper Res 40:764–775
25. Lee C-Y, Uzsoy R (1999) Minimizing makespan on a single batch processing machine with dynamic job arrivals. Int J Prod Res 37:219–236
26. Kempf K-G, Uzsoy R, Wang C-S (1998) Scheduling a single batch processing machine with secondary resource constraints. J Manuf Syst 17:37–51
27. Uzsoy R (1994) Scheduling a single batch processing machine with non-identical jobsizes. Int J Prod Res 32:1615–1635
28. Uzsoy R (1995) Scheduling batch processing machines with incompatible job families. Int J Prod Res 33:2685–2708
29. Mehta SV, Uzsoy R (1998) Minimizing total weighted tardiness on a batch processing machine with incompatible job families. IIE Trans 30:165–178
30. Devpura A, Fowler JW, Carlyle M, Perez I (2000) Minimizing total weighted tardiness on single batch process machine with incompatible job families. In: Proceedings of the symposium on operations research. Dresden, Germany, pp 366–371
31. Perez IC, Fowler JW, Carlyle WM (2005) Minimizing total weighted tardiness on a single batch process machine with compatible job families. Comput Oper Res 32:327–341
32. Wang C, Uzsoy R (2002) A genetic algorithm to minimize maximum lateness on a batch processing machine. Comput Oper Res 29:1621–1640
33. Melouk S, Damodaran P, Chang P-Y (2004) Minimizing make span for single machine batch processing with non-identical job sizes using simulated annealing. Int J Prod Econ 87:141–147
34. Koh S-G, Koo P-H, Kim D-C, Hur W-S (2005) Scheduling a single batch processing machine with arbitrary job sizes and incompatible job families. Int J Prod Econ 98:81–96
35. Kashan AH, Karimi B, Jenabi M (2008) A hybrid genetic heuristic for scheduling parallel batch processing machines with arbitrary job sizes. Comput Oper Res 35:1084–1098
36. Behnamian J, Zandieh M, Ghomi SMTF (2009) Parallel-machine scheduling problems with sequence-dependent setup times using an ACO, SA and VNS hybrid algorithm. Expert Syst Appl 36:9637–9644
37. Wang H-M, Chou F-D (2010) Solving the parallel batch-processing machines with different release times, job sizes, and capacity limits by metaheuristics. Expert Syst Appl 37:1510–1521

38. Fu Y, Tian G, Li Z, Wang Z (2018) Parallel machine scheduling with dynamic resource allocation via a master–slave genetic algorithm. IEEJ Trans Electr Electron Eng 13:748–756. https://doi.org/10.1002/tee.22625
39. Mohammadi G, Moaddabi E (2021) Using two metaheuristic algorithms for scheduling parallel machines with sequence dependent set-up times in job shop industries. Int J Syst Sci 52(14):2904–2917. https://doi.org/10.1080/00207721.2021.1913534

Chapter 5
Intelligent Integration of Process Planning and Scheduling

5.1 Introduction

Chryssolouris and Chan [1, 2] first proposed the idea of integrating process planning and scheduling (IPPS) problem in the mid-1980s. Meenakshi [3] and others emphasized that the integration of process planning and scheduling has great potential in improving productivity. According to the principle of human–computer interaction and collaborative work of different planners, Hitoshi [4] integrated process planning and scheduling into the production of different types of products in the flow workshop. Zhang [5] proposed a distributed integration model for the first time, which is different from the previous nonlinear or alternative process planning. This model is based on the available manufacturing resources in the workshop and the real-time feedback of workshop production. It mainly includes three modules: process planning module, production scheduling module and decision generation module.

Brandimarte and Calderini [6] proposed a hierarchical process planning and scheduling integration model, which considered that each job has multiple process routes, each process route is composed of multiple operations, and each operation can be processed by multiple machine tools. This model opens up new avenues for research on the integration of flexible process planning and shop-shop scheduling. The integrated method of process planning and scheduling based on collaborative mode proposed by Kempenaers [7] is based on the establishment of nonlinear process planning. It realizes real-time feedback of process planning and scheduling, thereby realizie dynamic process planning and scheduling. Saygin [8] discussed the importance of process planning and scheduling integration in flexible manufacturing systems, and proposed a flexible process planning method based on offline scheduling. In order to reduce the maximum completion time, this method adopts four strategies, namely flexible selection of machine tools, flexible selection of operations, scheduling and rescheduling. Kim [9] described the job shop scheduling problem of multiple parts with multiple flexible process plans. A process planning and scheduling model based on flexible job shops is constructed. The performance of

J. Wang, *Intelligent Manufacturing System and Intelligent Workshop*,
Advanced and Intelligent Manufacturing in China,
https://doi.org/10.1007/978-981-99-2011-2_5

pre-processing algorithms and iterative algorithms in solving process planning and scheduling integration is compared in his paper. Yang [10] proposed a multi-process route planning method based on part features and machine tool loads. According to the manufacturing features extracted from the part geometry, the machine tool capabilities, tooling equipment capabilities and process knowledge rules in the manufacturing resource library, the multi-process route is designed considering the load of the machine tool and processing time of each operation. The scheduling plan is optimized with the goal of delivery delay and lead time. Thomalla [11] proposed an optimization algorithm to solve the shop-floor scheduling problem based on alternative routes in a just-in-time (JIT) production environment. Wu [12] proposed a parallel integrated model of process planning and production scheduling in a distributed virtual manufacturing environment. This model uses multi-agent parallel computing and completes the optimal collaborative enterprise in a distributed environment through the value function. Kim [13] solves the IPPS problem through symbiotic evolutionary algorithm, adopts optimization strategies such as local crossover, replication, random symbiosis, combination selection and other genetic operators, and combines with traditional hierarchical algorithm and other co-evolutionary algorithms were compared. Moon [14] proposed an integrated model of process planning and scheduling among multiple factories. According to the data model, an evolutionary algorithm similar to genetic algorithm was used to solve the IPPS problem.

With the in-depth research on IPPS problems, a large number of technologies based on meta-heuristic algorithms have been applied to the research of this problem. In summary, they mainly include the following technologies:

(1) **Genetic Algorithm**

Lee [15] proposed a nonlinear process planning model based on genetic algorithm. Through the simulation module, each performance evaluation index of combined process planning is calculated, and these indexes are optimized by genetic algorithm to obtain a better scheduling plan. Moon [16] proposed an integrated model of process planning and scheduling based on multi-factory supply chain. With the minimum order delay as the optimization goal, the IPPS problem is solved through a heuristic algorithm based on genetic algorithm. Li has carried out fruitful research on genetic algorithm to solve IPPS problem [17–21]. Wong [22] modeled workshop dynamic information such as machine failures into the IPPS problem, and used genetic algorithms to solve the dynamic IPPS problem.

(2) **Ant Colony Algorithm**

There are few literatures on the application of ant colony algorithm to solve the IPPS problem. Kumar [23] tried to apply ant colony algorithm to solve the IPPS problem for the first time. Leung [24] proposed an integrated model of process planning and scheduling based on AND/OR diagrams, with the minimum completion time as the optimization goal, combining the ant colony search mechanism with multi-agents to solve process planning and scheduling problems. Wang [25] improved the IPPS problem model based on AND/OR graph, and used the ant colony algorithm to solve the IPPS problem. Leng [26] established a flexible process planning workshop scheduling model based on directed weighted graphs, and through dynamic adjustment of pheromone updates, it avoids the defects of slow convergence and local convergence of ant colony algorithm. On the basis of literature [24], Wong [27] improved the ant colony algorithm, and solved the process planning and scheduling problems through the two-stage ant colony algorithm, namely: two stages of operation selection and operation sequencing.

(3) **Particle swarm algorithm**

Guo [28] improved the particle swarm optimization algorithm, and designed several types of particle movement strategies for IPPS problems, which not only expanded the search space, but also effectively avoided local optimum. Guo [29] made further improvement on the basis of literature [28]. Aiming at the real-time nature of workshop production, a re-planning strategy was designed to cope with the dynamic changes of workshop production. Wang [30] applied the particle swarm optimization algorithm to multi-objective optimization of process planning and scheduling. The algorithm introduced a new solution expression scheme based on discrete search space. Sahraian [31] proposed a mixed integer linear programming model for the integration of process planning and scheduling among multiple factories. Dong [32] established an IPPS model based on multi-objective optimization and constraint mechanism for the characteristics of flexible job shops. In order to improve the search efficiency and avoid the occurrence of invalid solutions, an improved segmentation coding method was proposed.

In addition to the above three methods, methods including TS and mathematical programming are used to solve the IPPS problem. Kis [33] proposed a workshop scheduling system based on flexible technology, and applied TS algorithm to solve the workshop scheduling problem based on flexible technology. Tan [34] used mathematical programming method to solve the IPPS problem. Moreno [35] uses a hybrid evolutionary algorithm based on constraint satisfaction to solve planning and scheduling integration problems. Ueda [36] uses artificial neural network technology to solve the IPPS problems. Li [37] used simulated annealing algorithm to solve the IPPS problem. Mishra [38] proposed a hybrid algorithm of TS and SA based on chaos theory to solve the IPPS problems.

5.2 IPPS Problem Modelling

Process planning and scheduling have many commonalities in the manufacturing system. The common content of the process planning and scheduling must be considered comprehensively in order to achieve the optimization of integration of process planning and scheduling. Based on the content of the third chapter, the process planning is based on the part feature modelling technology, and the features of the part are used as the data source of the process planning. Since each feature of the part can be expressed by multiple process schemes, there are many kinds of process routes for each part, and there are differences in the number of operations and the sequence of operations. A scientific and reasonable scheduling plan can be formulated under the conditions of fully considering the utilization rate of manufacturing resources, the delivery date of production jobs and other factors. Therefore, the common factors of process planning and scheduling provide a theoretical basis for the integration.

So far, the research on IPPS is mainly based on three modes: nonlinear process planning (NLPP), closed-loop process planning (CLPP) and Just.in.time Process Planning (JTPP).

Nonlinear Process Planning (NLPP) model divides the IPPS problem into two consecutive stages. The first stage is process planning. First, it is assumed that the workshop is running smoothly, that is, the current manufacturing process is under controls during process planning. All feasible process routes for each part are formulated based on the current operating status of the manufacturing resources. In the second stage, workshop scheduling is carried out, and various process routes given a certain priority enter the scheduling system, and the scheduling system selects the optimal process route according to the specific resource conditions.

Closed-loop process planning (CLPP) is a dynamic process planning system oriented to workshop scheduling. It feeds back real-time information of workshop manufacturing resources to the process planning system, optimizes the process route in real time, and enables the process route to be dynamically modified and adjusted according to the actual production situation of the workshop. The production planning and production scheduling systems of the ERP system are integrated to form a closed loop for automatic adjustment of process routes. This dynamic system with automatic feedback function and integrated production plan, process planning and production scheduling can not only improve the real-time performance, but also improve the utilization rate of the process route of the process planning system.

Just.in.time Process Planning (JTPP) completes process planning and scheduling simultaneously. In this model process planning and scheduling is divided into two stages, the preliminary planning stage and the detailed planning stage. The preliminary planning stage is mainly to analyze production tasks, including information such as batch size, production type, part features, and the relationship between features and features, and determine the preliminary process route according to the production type and feature information of parts; in the detailed planning stage, the workshop processing equipment information is matched with the production task information, and a complete process route and scheduling plan are generated at the same time.

The research method adopted in this chapter is based on the idea of NLPP integration model. The advantage of this model is that it generates all possible process routes, which expand the optimization space of workshop scheduling and help to find the optimal process route; its disadvantage is that it will face the combinatorial explosion problem when the scale of the NLPP integration model is large, which would result in the difficulty and time of solving are greatly increased. The description of the NLPP model on the IPPS problem is as follows.

The n jobs $\{J_1, J_2, \ldots, J_n\}$ are processed on m machine tools $\{W_1, W_2, \ldots, W_m\}$. Each job consists of at most p operations $\{O_{i1}, O_{i2}, \ldots, O_{ip}\}$, and form q optional process routes $\{L_{i1}, L_{i2}, \ldots, L_{iq}\}$ composed of p operations according to certain rules. The determined q optional process route $\{L_{i1}, L_{i2}, \ldots, L_{iq}\}$ are input into the scheduling system as the data source. The scheduling system determine the process routes of each part according to the scheduling strategy of the, and output the optimal scheduling scheme. The assumptions described in Chap. 4 on flexible job shop scheduling are still needed here.

The optimization goal of the IPPS problem is based on workshop manufacturing resources and production efficiency. Therefore, the evaluation indicators of the IPPS problem are the same as the FJSP.

Therefore, the optimization strategy is mainly divided into two stages:

The first stage: According to different optimization objectives, multiple process routes for different part are determined respectively.

The second stage: According to the multiple process routes of each part determined in the first stage, the workshop resource scheduling is carried out.

The IPPS optimization strategy based on the NLPP integration model is shown in Fig. 5.1.

IPPS problem can presented using AND/OR graph, which is shown in Fig. 5.2.

Figure 5.2 shows alternative process routes for two parts. The graph mainly includes three elements: nodes, directed arcs, and AND/OR relationships. The node represents the operations of the part, the directed arc represents the sequence between the operations, and AND/OR represents the relationship among some operations. Part

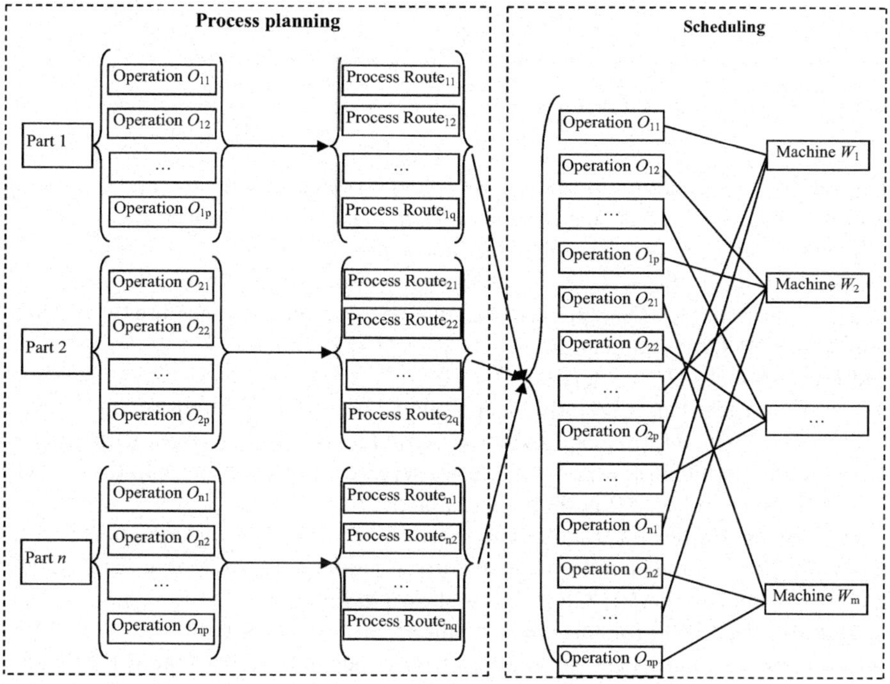

Fig. 5.1 Optimization strategy for IPPS

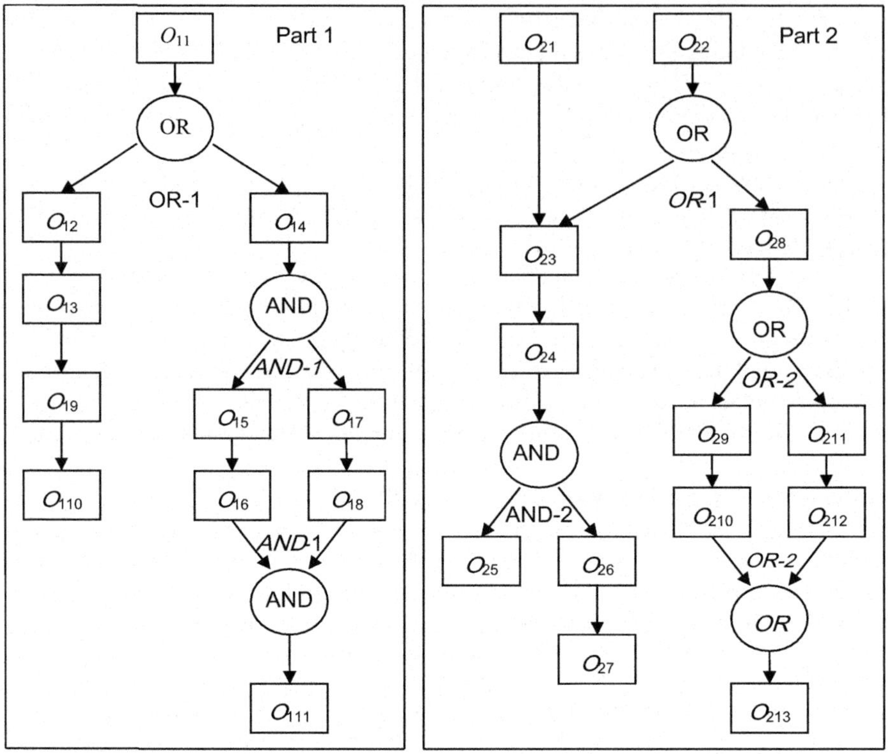

Fig. 5.2 AND/OR graph of Part 1 and Part 2

1 in the Fig. 5.2 contains 11 operations. The first operation is O_{11}, and the second operation has two options, that is, either O_{12} or O_{14}. Because there is an OR relationship, you cannot choose O_{14} when you choose O_{12}. If operation O_{12} is selected, $O_{11} \rightarrow O_{12} \rightarrow O_{13} \rightarrow O_{19} \rightarrow O_{110}$ represents the optional process route OL_{11} of Part 1. If O_{14} is selected, O_{15}, O_{16}, O_{17}, and O_{18} must be selected. So, there are two optional process routes OL_{12}, namely $O_{11} \rightarrow O_{14} \rightarrow O_{15} \rightarrow O_{16} \rightarrow O_{17} \rightarrow O_{18} \rightarrow O_{111}$, and OL_{13}, namely $O_{11} \rightarrow O_{14} \rightarrow O_{17} \rightarrow O_{18} \rightarrow O_{15} \rightarrow O_{16} \rightarrow O_{111}$. Therefore, it can be seen that Part 1 include 11 operations, one AND relationship and one OR relationships, which form 3 optional process routes OL_{11}, OL_{12}, OL_{13}. Part 2 include 13 operations, which form 6 optional process routes. The process route for the IPPS problem shown in Fig. 5.2 is shown in Table 5.1.

It is necessary to add the alternative machine tool for each operation and the processing time of the operation on the corresponding machine tool in the AND/OR graph. The improved AND/OR graph is shown in Fig. 5.3.

Because the IPPS problem is a scheduling problem of a multi-part flexible job shop with uncertain process routes, the performance indicators of the IPPS problem is basically similar to the FJSP. The performance indicators and their definitions are shown in Sect. 4.2.1.

Table 5.1 Process route for the IPPS Problem in Fig. 5.2

Part	No	Process route	Content
Part 1	1	OL_{11}	$O_{11} \to O_{12} \to O_{13} \to O_{19} \to O_{110}$
	2	OL_{12}	$O_{11} \to O_{14} \to O_{15} \to O_{16} \to O_{17} \to O_{18} \to O_{11}$
	3	OL_{13}	$O_{11} \to O_{14} \to O_{17} \to O_{18} \to O_{15} \to O_{16} \to O_{111}$
Part 2	1	OL_{21}	$O_{21} \to O_{23} \to O_{24} \to O_{25} \to O_{26} \to O_{27}$
	2	OL_{22}	$O_{21} \to O_{23} \to O_{24} \to O_{26} \to O_{27} \to O_{25}$
	3	OL_{23}	$O_{22} \to O_{23} \to O_{24} \to O_{25} \to O_{26} \to O_{27}$
	4	OL_{24}	$O_{22} \to O_{23} \to O_{24} \to O_{26} \to O_{27} \to O_{25}$
	5	OL_{25}	$O_{22} \to O_{28} \to O_{29} \to O_{210} \to O_{213}$
	6	OL_{26}	$O_{22} \to O_{28} \to O_{21} \to O_{212} \to O_{213}$

5.3 Using Genetic Algorithm to Solve the IPPS Problem

1. **Gene Encoding**

According to the improved AND/OR graph, it can be seen that the IPPS problem is actually an integration of flexible process planning and flexible job shop scheduling. On this basis, the chromosome is composed of three parts, the process route gene, the operation gene and the machine tool gene. The part of operation gene and the machine tool gene refer to the chromosome coding scheme of the FJSP in Chapter Four. The part of process route gene is designed separately. Since the IPPS problem not only needs to form the process route of each part, but also forms the final scheduling plan. Therefore, referring to the coding scheme of the FJSP, a chromosome coding scheme with the operations as the basic information is formed. Chromosome A for the IPPS problem of Fig. 5.3 is shown in Table 5.2.

The process route gene part represents the process route codes of all parts in the IPPS problem, its code digits are the number of parts involved in the IPPS problem, and the value of each code position is the process route code of each part. In this example, the process route gene part contains two code points, which represent the process route codes of Part 1 and Part 2 respectively. According to Table 5.1, the alternative codes on the 1th code position are {1, 2, 3}. The alternative codes on the 2th code position are {1, 2, 3, 4, 5, 6}. In this example, the '3' on the 1th code position of the process route gene part represents the third process route of Part 1, and the '2' in the second code position represents the 2th process route of Part 2.

The number of code position contained in the operation gene part is the number of all operations contained in the selected process route of all parts in the IPPS problem. For chromosome A, because the 3th process route of Part A contains 7 operations, the 2th process route of Part 2 contains 6 operations. Therefore, the operation gene part of chromosome A contains 13 code points. The value of each code point is the number of the part, that is, all 7 operations of Part A are 1, and all 6 operations of Part B are 2. So, the operations gene of the entire chromosome consists of 7 numbers '1' and 6 numbers '2'.

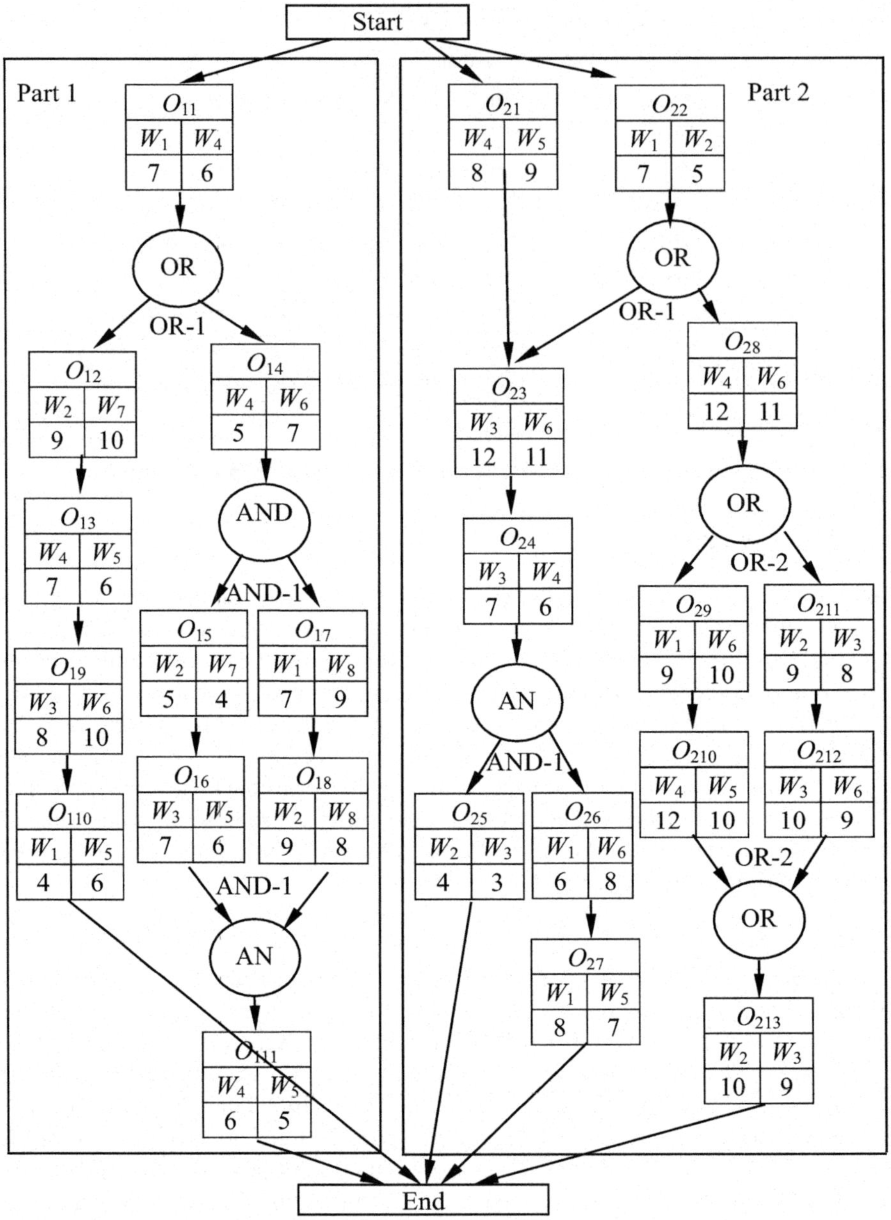

Fig. 5.3 The improved AND/OR graph

Table 5.2 Chromosome A

Code	Gene part	O_{11}	O_{21}	O_{14}	O_{17}	O_{23}	O_{24}	O_{18}	O_{26}	O_{15}	O_{27}	O_{25}	O_{16}	O_{111}
OL	Process routes	3	2											
OP	Operations	1	2	1	1	2	2	1	2	1	2	2	1	1
M	Machine tools	1	4	6	8	6	3	2	1	2	5	2	3	5

The machine tool gene part is the machine tool code of the corresponding operation represented by the operation gene part. For example, number '1' on the 1th code position of the operation gene part represent the operation O_{11} in the 3th process route of Part 1, and the corresponding code position of the machine tool gene part is '1'. According to Fig. 5.3, it can be found that this operation can be obtained by two machine tools W_1 and W_4, and the processing time is 7 s and 6 s respectively. Similarly, the 11th code position of the operation gene part is '2', which represents the operation O_{25}. The corresponding machine tool code position is '2', which means that the processing machine tool of this process is W_2.

2. **Initializing population**

Because the chromosome is composed of 3 gene parts, the chromosomes are initialized separately according to the different coding rules. The initialization process is as follows:

① According to the IPPS problem, generate the process route code table T_l, operation code table T_o, and machine tool code table T_m of each part.
② Update the operation code table T_o according to the AND/OR graph of the IPPS problem.
③ According to the process route code table, randomly generate a random number from 1 to n (the number of process routes for each part is n) to form the process route gene part.
④ According to the process route selected in step ③ and the operation code table T_o, randomly generate the operation gene part.
⑤ According to the operation gene part in step ④ and the machine tool code table T_m generated, randomly generate the machine tool gene part.

Among them, the operation code table T_o and the process route code table T_l are shown in Table 5.3 and Table 5.4 respectively. The process route formed in the iteration need to record access operation dynamically, which is shown in Table 5.5.

3. **Copy, Crossover and Mutation**

Similar to solving the CAPP and FJSP problems, the tournament selection method is adopted to perform copy operation. To avoid falling into a local optimum, the copy probability for chromosomes with the same or close fitness value in elite chromosomes (chromosomes with the highest fitness in the population) is set to be 10–20%.

Table 5.3 Operation code table T_o

Data types	Variables	Description
Int	Op_id	Number of operations
Int	M_id[]	Candidate machine tools of operations set
Dec	M_t[]	Process time set
Int	Op_id[]	Candidate operations set
Int	Op_tp[]	Visit type, −1: OR, 0: None, 1: AND

Table 5.4 Process route code table T_l

Data types	Variables	Description
Int	Pt_id	Number of parts
Int	Ol_id	Number of process routes
Int	Op_id[]	Operations set

Table 5.5 Visiting operation code table

Data types	Variables	Description
Int	Op_id	Number of operations
Int	M_id	Number of machine tool
Dec	M_t	Process time
Int()	Op_id_next	Candidate operation set

This set not only retains elite chromosomes in the next generation population, but also voids GA falling into local optimum.

In order to explain the crossover and mutation operations more concisely and clearly, the IPPS problem composed of two parts in Fig. 5.3 is used as an example. The chromosome B constructed is shown in Table 5.6.

Because the gene coding include three parts for the IPPS problem, that is, the process route genetic part, the operation genetic part and the machine tool genetic part, the crossover operation is designed for the three parts respectively.

Table 5.6 Chromosome B

Code	Gene part	O_{11}	O_{21}	O_{14}	O_{15}	O_{16}	O_{17}	O_{18}	O_{23}	O_{24}	O_{26}	O_{27}	O_{111}	O_{25}
OL	Process routes	3	2											
OP	Operations	1	2	1	1	1	1	1	2	2	2	2	1	2
M	Machine tools	4	5	4	2	3	8	8	6	4	1	5	5	3

Process route gene part

Because the chromosome coding is based on the operation information, it must be ensured that the two chromosome schemes selected have the same process and machine code position. For the two chromosomes selected for the crossover operation, each code position of the process route gene part cannot be determined arbitrarily. Therefore, the crossover operation process of the process route gene part is as follows:

Step 1: If the number N of parts in the IPPS problem is less than 3, the crossover operation of the process route gene part is not performed, otherwise, go to Step 2.

Step 2: Randomly select the chromosome A for the crossover operation, determine the process route number of each part contained in the chromosome, and check whether the variable Op_tp [] of each operation is equal to 1, if not, a chromosome B is selected with probability, which must have the same number of process route gene parts as chromosome A. If Op_tp [] is equal to 1, go to STEP 3.

Step 3: Select other chromosome B with the same AND relation as chromosome A with a certain probability.

Operation gene part

Step 1: Randomly generate two intersection points p, q.

Step 2: For the chromosome A, take out the operation genes between the intersection points p and q of, and keep the operation genes outside the intersection points unchanged.

Step 3: Find the missing operation genes of the chromosome A, and insert into the position between the intersection points of the chromosome A according to the original arrangement order of the chromosome B. Then, the operation gene part of the offspring chromosome C forms.

Step 4: The machine tool genes of the new operation gene part of the chromosome C are filled with the machine tool selected by the corresponding operation in the chromosome A.

Take chromosome A in Table 5.5 and chromosome B in Table 5.6 as examples to illustrate the process of the two-point crossover. The intersection points of chromosome A are taken as gene points 4 and 8. The process of two points crossover on operation gene is shown in Fig. 5.4, and the new offspring chromosome C is shown in Table 5.7.

Similarly, the chromosome D based on the crossover of the chromosome B can be generated according to the above crossover rules, as shown in Table 5.8.

Machine tool gene part

Due to the limitation of the processing machines allowed by the operations, the use of two-point crossover on the machine tool genetic part may cause the failure of the crossover operation. The machine tool gene part uses a uniform crossover operation. The workflow of the crossover operation on machine tool gene part is as follows:

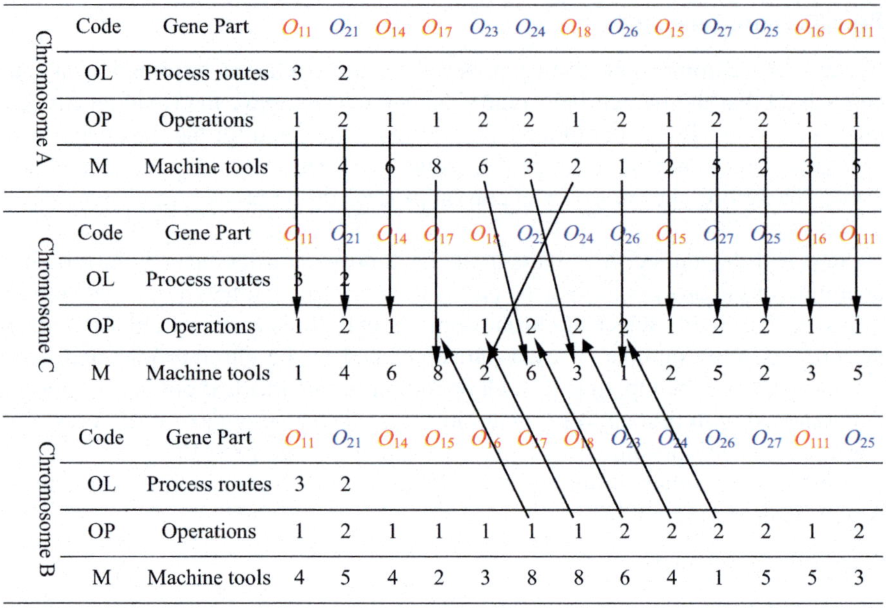

Fig. 5.4 Two points crossover on operation gene

Table 5.7 Chromosome C

Code	Gene part	O_{11}	O_{21}	O_{14}	O_{17}	O_{18}	O_{23}	O_{24}	O_{26}	O_{15}	O_{27}	O_{25}	O_{16}	O_{111}
OL	Process routes	3	2											
OP	Operations	1	2	1	1	1	2	2	2	1	2	2	1	1
M	Machine tools	1	4	6	8	2	6	3	1	2	5	2	3	5

Table 5.8 Chromosome D

Code	Gene part	O_{11}	O_{21}	O_{14}	O_{17}	O_{23}	O_{18}	O_{15}	O_{16}	O_{24}	O_{26}	O_{27}	O_{111}	O_{25}
OL	Process routes	3	2											
OP	Operations	1	2	1	1	2	1	1	1	2	2	2	1	2
M	Machine tools	4	5	4	8	6	2	2	3	4	1	5	5	3

Step 1: Select the above chromosomes C and D generated by performing the crossover operation on the operation gene part.

Step 2: Randomly generate two intersection points p, q, which represents the number of the process route.

Table 5.9 Chromosome E

Code	Gene part	O_{11}	O_{21}	O_{14}	O_{17}	O_{18}	O_{23}	O_{24}	O_{26}	O_{15}	O_{27}	O_{25}	O_{16}	O_{111}
OL	Process routes	3	2											
OP	Operations	1	2	1	1	1	2	2	2	1	2	2	1	1
M	Machine tools	4	5	4	8	2	6	4	1	2	5	3	3	5

Step 3: Keep the operation genetic part of chromosome C unchanged, cross the machine tools of all operations of process routes *p* and *q* in chromosome C and chromosome D.

Take chromosome C in Table 5.7 and chromosome D in Table 5.8 as examples to perform crossover operation. For this example, because there are only two parts, the intersection positions are only 3 and 2. The chromosome C performs crossover to form a offspring chromosome E shown in the Table 5.9. The execution process is shown in Fig. 5.5. Similarly, the chromosome D performs crossover to form a offspring chromosome F, as shown in Table 5.10.

Similarly, the mutation operation is designed for the process route gene part, the operation gene part and the machine tool gene part respectively.

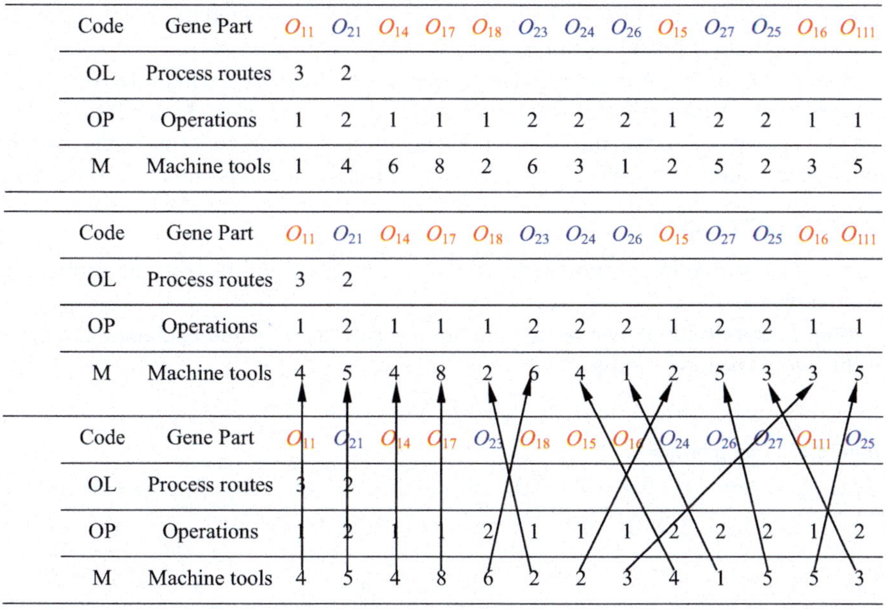

Fig. 5.5 Uniform crossover on machine tools gene

Table 5.10 Chromosome F

Code	Gene part	O_{11}	O_{21}	O_{14}	O_{17}	O_{23}	O_{18}	O_{15}	O_{16}	O_{24}	O_{26}	O_{27}	O_{111}	O_{25}
OL	Process routes	3	2											
OP	Operations	1	2	1	1	2	1	1	1	2	2	2	1	2
M	Machine tools	1	4	6	8	6	2	2	3	3	1	5	5	2

Process route gene part

Step 1: If the number N of parts in the IPPS problem is less than 2, the mutation operation of the process route gene part is not performed, otherwise, go to Step 2.

Step 2: Select the chromosome A for mutation operation with a certain probability, randomly select two points p and q for mutation operation, and represent the number of the part that is mutated.

Step 3: Keep the operation genes parts that do not mutate in chromosome A unchanged. The operations from in the new process route generated from the mutation are added to chromosome A according to the original operation gene. If there are not enough vacancies, their operations are arranged to the bottom of the chromosome in sequence.

Step 4: Keep The machine tool genes of the parts that do not mutate in chromosome A unchanged. The machine tool number is randomly determined for those operations from the mutation in chromosome A.

Take chromosome A in Table 5.2 as an example to illustrate the mutation operation. Chromosome A has only two parts, so p and q are 1 and 2 respectively. Assume that the first code position and the second code position of the process route gene part are mutated to 2 and 4 respectively, and the new chromosome G is shown in Table 5.11.

Operation gene part

Step 1: Determine the chromosome A for mutation operation with the probability set in advance.

Step 2: Randomly generate two mutation positions p, q, and exchange the genes at the two positions p and q.

Table 5.11 Chromosome G

Code	Gene part	O_{11}	O_{22}	O_{14}	O_{15}	O_{23}	O_{24}	O_{16}	O_{26}	O_{17}	O_{27}	O_{25}	O_{18}	O_{111}
OL	Process routes	2	4											
OP	Operations	1	2	1	1	2	2	1	2	1	2	2	1	1
M	Machine tools	1	2	6	7	6	4	3	1	1	5	2	2	4

Table 5.12 Chromosome H

Code	Gene part	O_{11}	O_{21}	O_{14}	O_{17}	O_{23}	O_{24}	O_{18}	O_{26}	O_{15}	O_{27}	O_{111}	O_{16}	O_{25}
OL	Process routes	3	2											
OP	Operations	1	2	1	1	2	2	1	2	1	2	2	1	1
M	Machine tools	1	4	6	8	6	3	2	1	2	5	5	3	3

Step 3: Check whether the operation gene part meets the requirements, that is, the latter operation can only be carried out after the previous operation is finished. If the requirements are not met, go to Step 2.

Step 4: Exchange the corresponding machine tools of operations from the two mutation p and q.

For the chromosome A in Table 5.2, take the mutation points as 3 and 8. When the corresponding operation O_{14} and operation O_{26} are exchanged, Operation O_{17} and operation O_{18} is arranged before operation $O_{14,}$ and operation O_{26} is arranged before operation O_{23} and operation O_{24}, which does not meet the actual process constraints obviously. Therefore, taking the mutation points as 11 and 13, the offspring chromosome H obtained through the mutation operation is shown in Table 5.12.

Machine tool gene part

Step 1: Randomly select the chromosomes H generated by the mutation operation on the operation gene part.

Step 2: Randomly generate two mutation points p, q, which represent the number of the machine tool to be mutated.

Step 3: Keep the operation gene part unchanged, and the machine tool of the operation on the mutation points p and q is replaced to be other machine tool numbers of the same operation.

Therefore, for chromosome H in Table 5.12, take the mutation points as 3 and 8. The machine tools of operation O_{14} and process O_{26} are changed from machine tool 6 and machine tool 1 to machine tool 4 and machine tool 6 respectively, and the offspring chromosome I after mutation is shown in Table 5.13.

Table 5.13 Chromosome I

Code	Gene part	O_{11}	O_{21}	O_{14}	O_{17}	O_{23}	O_{24}	O_{18}	O_{26}	O_{15}	O_{27}	O_{111}	O_{16}	O_{25}
OL	Process routes	3	2											
OP	Operations	1	2	1	1	2	2	1	2	1	2	2	1	1
M	Machine tools	1	4	4	8	6	3	2	6	2	5	5	3	3

References

1. Chryssolouris G, Chan S, Cobb W (1984) Decision making on the factory floor: an integrated approach to process planning and scheduling. Robot Comput Integr Manuf 1(3–4):315–319
2. Chryssolouris G, Chan S (1985) An integrated approach to process planning and scheduling. Ann CIRP 34(1):413–417
3. Meenakshi Sundaram R, Fu SS (1988) Process planning and scheduling—a method of integration for productivity improvement. Comput Ind Eng 15:296–301
4. Tsubone H, Anzai M, Sugawara M et al (1991) Interactive production planning and scheduling system for a flow-type manufacturing process. Int J Prod Econ 22(1):43–51
5. Zhang HC (1993) IPPM-A prototype to integrated process planning and job shop scheduling functions. Ann CIRP 42(1):513–518
6. Brandimarte P, Calderini M (1995) A hierarchical bicriterion approach to integrated process plan selection and job shopscheduling. Int J Prod Res 33:161–181
7. Kempenaers J, Pinte J, Detand J, Kruth JP (1996) Collaborative process planning and scheduling system. Adv Eng Softw 25(1):3–8
8. Saygin C, Kilic SE (1999) Integrating flexible process plans with scheduling in flexible manufacturing systems. Int J Adv Manuf Technol 15(4):268–280
9. Kim KH, Egbelu PJ (1999) Scheduling in a production environment with multiple process plans per job. Int J Prod Res 37:2725–2753
10. Yang YN, Parsaei HR, Leep HR (2001) A prototype of a feature-based multiple-alternative process planning system with scheduling verification. Comput Ind Eng 39:109–124
11. Thomalla C (2001) Job shop scheduling with alternative process plans. Int J Prod Econ 74:125–134
12. Wu SH, Fuh JYH, Nee AYC (2002) Concurrent process planning and scheduling in distributed virtual manufacturing. IIE Trans 34:77–89
13. Kim Y, Park K, Ko J (2003) A symbiotic evolutionary algorithm for the integration of process planning and job shop scheduling. Comput Oper Res 30:1151–1171
14. Moon C, Seo Y (2005) Evolutionary algorithm for advanced process planning and scheduling in a multi-plant. Comput Ind Eng 48:311–325
15. Lee H, Kim S (2001) Integration of process planning and scheduling using simulation based genetic algorithms. Int J Adv Manuf Technol 18:586–590
16. Moon C, Kim J, Hur S (2002) Integrated process planning and scheduling with minimizing total tardiness in multi-plants supply chain. Comput Ind Eng 43:331–349
17. Li XY, Zhang ChY, Gao L, Shao XY (2010) An agent-based approach forintegrated process planning and scheduling. Expert Syst Appl 37(2):1256–1264
18. Li XY, Gao L, Shao XY, Zhang CY (2010) Mathematical modeling and evolutionary algorithm-based approach for integrated process planning and scheduling. Comput Oper Res 37(4):656–667
19. Li XY, Shao XY, Gao L, Zhang ChY (2010) An effective hybrid algorithm for integration of process planning and scheduling. Int J Prod Econ 126:289–298
20. Li X, Gao L, Shao X (2012) An active learning genetic algorithm for integrated process planning and scheduling. Expert Syst Appl 39(8):6683–6691
21. Xia H, Li X, Gao L (2016) A hybrid genetic algorithm with variable neighborhood search for dynamic integrated process planning and scheduling. Comput Comput Ind Eng 102(1):99–112
22. Zhang LP, Wong TN (2015) An object-coding genetic algorithm for integrated process planning and scheduling. Eur J Oper Res 244:434–444
23. Kumar R, Tiwari MK, Shankar R (2003) Scheduling of flexible manufacturing systems: an antcolony optimization approach. Proc Inst Mech Eng Part B J Eng Manuf 217(10):1443–1453
24. Leung CW, Wong TN, Mak KL, Fung RYK (2010) Integrated process planning and scheduling by anagent-based ant colony optimization. Comput Ind Eng 59(1):166–180
25. Wang JF, Fang XL, Zhang CW et al (2014) A graph-based Ant colony optimization approach for integrated process. Plann Sched 22:748–753

26. Leng S, Wei XB, Zhang WY (2006) Improved ACO scheduling algorithm based on flexible process. Trans Nanjing Univ Aeronaut Astronaut 23(2):154–159
27. Wong TN, Zhang S, Wang G (2012) Integrated process planning and scheduling—multi-agent system with two-stage ant colony optimisation algorithm. Int J Prod Res 50(21):6188–6201
28. Guo YW, Li WD, Mileham AR, Owen GW (2009) Applications of particle swarm optimization in integrated process planning and scheduling. Robot Comput Integr Manuf 25:280–288
29. Guo YW, Li WD, Mileham AR, Owen GW (2009) Optimisation of integrated process planning and scheduling using a particle swarm optimisation approach. Int J Prod Res 47(4):3775–3796
30. Wang YF, Zhang YF, Fuh JYH (2010) A PSO-based multi-objective optimization approach to the integration of process planning and scheduling. In: Proceeding of 8th IEEE international conference on control and automation. China, pp 214–219
31. Sahraian R, Haghighi AK, Ghasemi E (2011) A multi-objective optimization model to the integrating flexible process planning and scheduling based on modified particle swarm optimization algorithm (MPSO). World Acad Sci Eng Technol 79:85–692
32. Dong QY, Lu JSh, Gui YK (2012) Integrated optimization of production planning and scheduling for flexible job-shop. Int Rev Comput Softw 7(3):1273–1282
33. Kis T (2003) Job shop scheduling with processing alternatives. Eur J Oper Res 151:307–332
34. Tan W, Khoshnevis B (2004) A linearized polynomial mixed integer programming model for the integration of process planning and scheduling. J Intell Manuf 15:593–605
35. Moreno MDR, Oddi A, Borrajo D (2006) IPSS: a hybrid approach to planning and scheduling integration. IEEE Trans Knowl Data Eng 18(12):1681–1695
36. Ueda K, Fuji N, Inoue R (2007) An emergent synthesis approach to simultaneous process planning and scheduling. Ann CIRP 56(1):463–466
37. Li WD, McMahon C (2007) A simulated annealing-based optimization approach for integrated process planning and scheduling. Int J Comput Integr Manuf 20(1):80–95
38. Mishra N, Choudhary AK, Tiwari MK (2008) Modeling the planning and scheduling across the outsourcing supply chain: a chaos-based fast Tabu-SA approach. Int J Prod Res 46(13):3683–3715

Chapter 6
Case of Intelligent Manufacturing

6.1 Aerospace Smart Factory Practice

6.1.1 Aerospace Intelligent Manufacturing System Architecture

The development of the aerospace industry is important to national security strategy, and represents the country's scientific and technological ability to enter and exit space independently. As basic space transportation vehicles and space carriers, Spacecraft products such as launch vehicles, Shenzhou spacecraft, and space stations are the basic guarantee for the realization of all space programs.

An Aerospace Equipment Manufacturing Factory undertakes the military, meteorological, marine, geological, communication and international commercial launch missions. Due to different payloads, space orbits, etc., the structural parts of each spacecraft will change. So most of the aerospace products are the customized products. Faced with the new situation of high-density space launch and rapid access to space, there are many increasingly prominent problems such as long development cycle, low efficiency, high cost, and low quality resulted from multi-model cross-parallel production and unbalanced production volume.

1. **Manufacturing characteristics of aerospace products**

The manufacturing technology of aerospace products has the features of advancement, complexity, integration and extreme manufacturing. These technical characteristics determine that it will inevitably develop in the direction of intelligent manufacturing. Some of the above-mentioned features of aerospace manufacturing make it necessary to widely use automatic control and information technology to further improve the level of technology and reliability based on the existing material technology, design technology and process technology. Aerospace manufacturing has the characteristics of small-batch multi-variety mixed production, higher quality

J. Wang, *Intelligent Manufacturing System and Intelligent Workshop*,
Advanced and Intelligent Manufacturing in China,
https://doi.org/10.1007/978-981-99-2011-2_6

and reliability requirements, and unbalanced development of production and technology. These characteristics bring some problems such as unreasonable utilization of resources and long development cycle.

To sum up, due to the characteristics of aerospace products, the application of intelligent manufacturing has become an inevitable trend.

2. **Aerospace product manufacturing system architecture**

According to the characteristics of spacecraft structural parts, such as single piece, small batch and distributed, an overall framework of the intelligent manufacturing workshop of spacecraft structural parts is established, as shown in Fig. 6.1.

Intelligent process design and real-time information perception and feedback are realized through the development of key intelligent features such as intelligent numerical control system and intelligent servo drive, which could meet the efficient and reliable manufacturing of spacecraft structural parts.

3. **General idea of aerospace intelligent manufacturing**

Facing the production and technological characteristics of spacecraft structural parts, the overall technical route is to realize the overall design and layout simulation of the intelligent workshop of the spacecraft structural parts through the correspondence, fusion, cycle optimization and improvement of the physical space and virtual space of the workshop, which is shown in Fig. 6.2. The collaborative integration of information flow, material flow and business flow is realized based on the design and plan of human, machine, material, method, environment, and measurement in the physical space. Intelligent workshops are constructed based on CPS. The intelligent workshop is composed of several intelligent units. According to the manufacturing task requirements of the spacecraft structural parts, the process planning is carried out. The equipment in the intelligent units is combined into the production line to realize the flexible layout method. A virtual workshop consistent with the physical workshop is constructed by modeling the human, machine, material, ring and other equipment resources in the physical space on the software platform. All production business activities in the physical workshop are planned, evaluated and verified based on the virtual workshop. Through modeling and simulation, the uncertainty in the transformation from design to implementation of workshop and production line is reduced. The success rate and reliability of the intelligent manufacturing system of spacecraft structural parts are improved.

At the same time, the real data of the logistics workshop fed back by the CPS system is brought into the digital model of the virtual workshop to realize the three-dimensional visualization of the real-time data of the manufacturing execution process. The planning and layout of the physical workshop are guided by the digital model of the virtual workshop by the feeding back real-time production data.

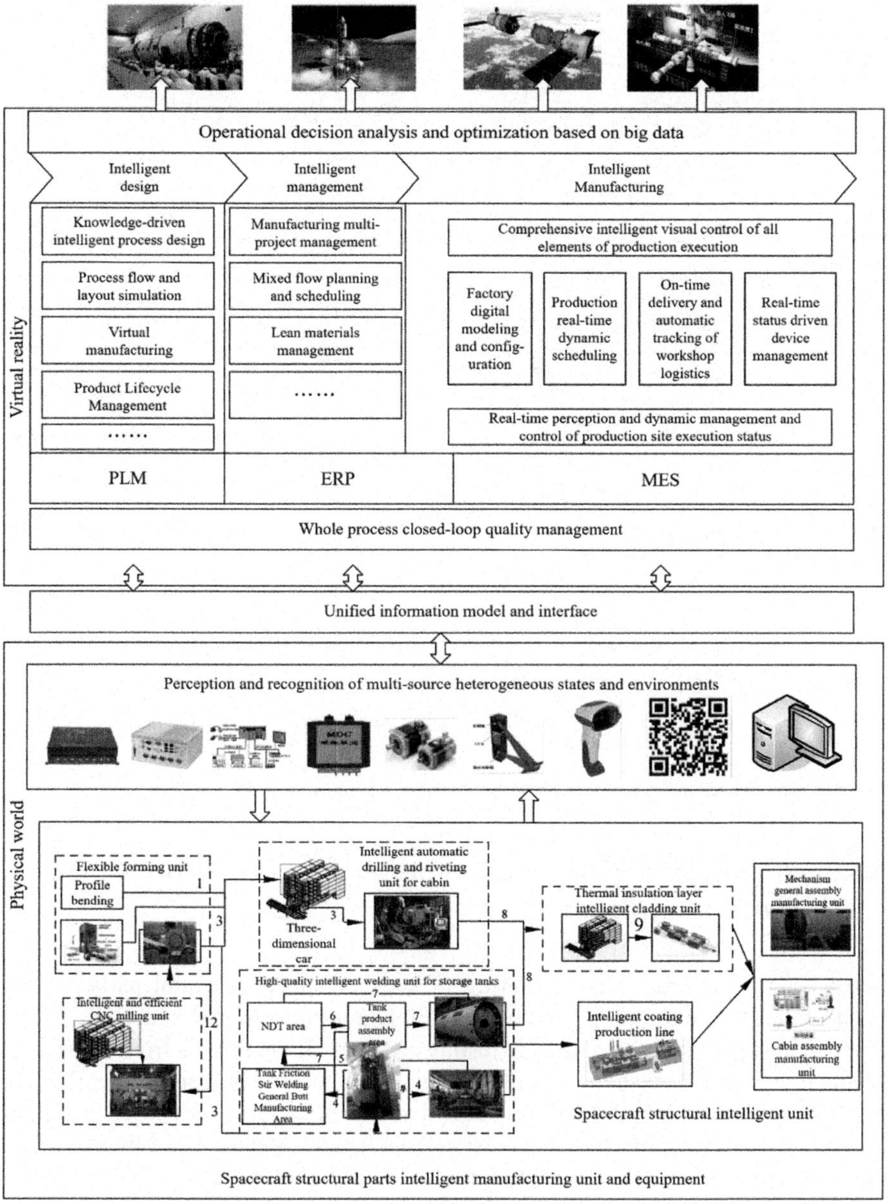

Fig. 6.1 Aerospace Production System architecture

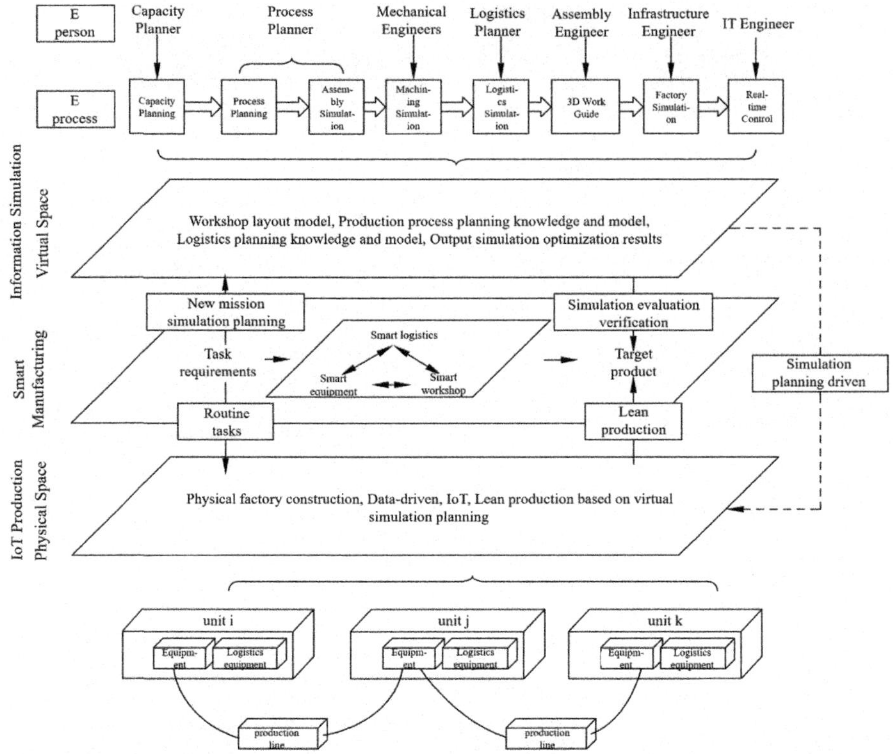

Fig. 6.2 The overall design and layout simulation of the smart workshop

6.1.2 *Industrial IoT Application in Aerospace Structural Parts Manufacturing*

The basis of intelligent manufacturing of key aerospace structural parts is interconnection. It is reflected in the horizontal integration from raw materials, design, process, manufacturing, assembly to testing and vertical integration from equipment, production lines, workshops to enterprises. The accounts of aerospace structural parts are more than 70% of the total aerospace manufacturing. In order to support the aerospace development strategy, the key application standards for the intelligent manufacturing of aerospace structural parts should be established.

Firstly, improve the intelligent manufacturing standard system According to the intelligent manufacturing standardization construction guidelines, the standards established are linked together. The design, equipment and technology of the manufacturing process of key structural parts are integrated with the identification of the in-process product as the core.

Secondly, promote the implementation of intelligent manufacturing in aerospace manufacturing Industrial Internet of Things technology is applied in the

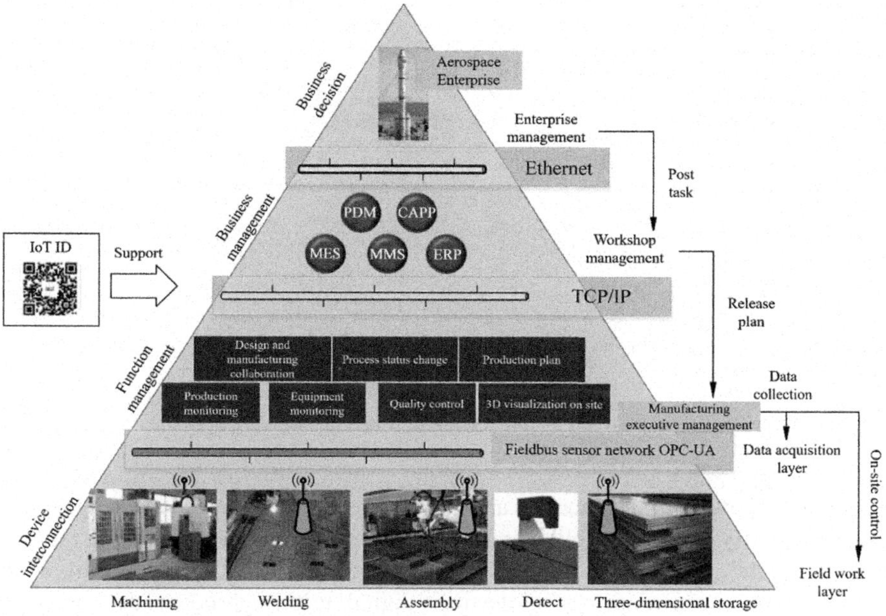

Fig. 6.3 Vertical integration of manufacturing equipment and systems

intelligent manufacturing of key aerospace structural parts to improve manufacturing efficiency, quality control and full life cycle traceability.

1. **Aerospace Intelligent Manufacturing Interconnection Model**

The overall effect of aerospace intelligent manufacturing is flexible, transparent, controllable, digital, and intelligent. The interconnection between manufacturing equipment and manufacturing systems of key aerospace structural parts is mainly reflected in the vertical integration of equipment and systems, as shown in Fig. 6.3.

According to the manufacturing process of the key structural parts of aerospace products, the layout of the simulated manufacturing workshop is formulated. An industrial IoT system for key aerospace structural parts is established based on the integration of connection of the upstream and downstream of the production process. The manufacturing process of key structural parts is simply divided into five stations: welding, machining, heat treatment, assembly and testing. The virtual workshop also includes machining centers, three-dimensional warehouses, materials, labor consoles, AGV trolleys and their tracks. The verification relationship on the structural parts between the overall process and the standard is shown in Fig. 6.4.

2. **Implementation steps of aerospace intelligent manufacturing**

The standards related to intelligent manufacturing in the aerospace industry focus on the information topics in the group-level standard system, covering terminology

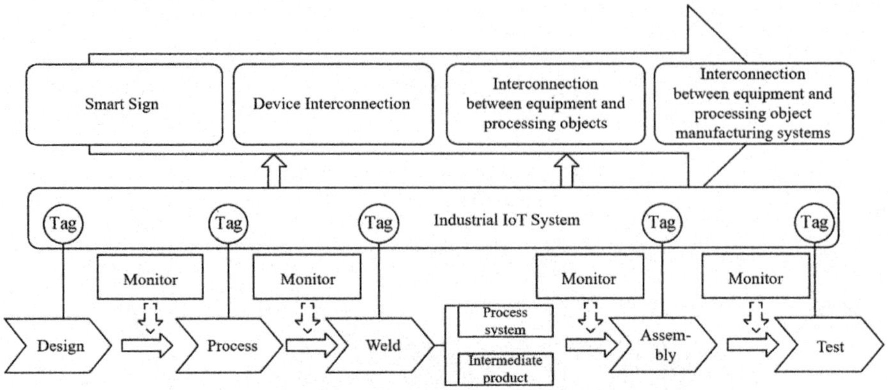

Fig. 6.4 The relationship between the processing flow and the standard verification

symbols, information classification and coding, data expression and exchange, information security, digital modeling and definition, digital pre-assembly, Digital technology, digital processing, digital manufacturing execution, manufacturing simulation, digital inspection, digital testing, digital testing, product data management, supply chain management, enterprise resource planning, etc.. These standards provide standardized support for aerospace digital design, manufacturing and production.

The following standards are mainly used in an aerospace smart factory from the perspective of process and information integration.

The digital process design and simulation standard group includes design process coordination standards, process design and management standards, process simulation standards, and process basic standards.

The digital integrated manufacturing standard group includes digital factory/workshop/production line/unit construction standards, production scheduling standards, logistics management standards, and quality management standards.

The industrial IoT standard. Industrial Internet of Things reference model, general requirements of intelligent identification, specification requirements for interconnection between manufacturing equipment, technical framework for interconnection between manufacturing equipment and processing objects, technical requirements for interconnection between manufacturing equipment and manufacturing systems.

On the basis of the new mode of intelligent manufacturing of aerospace products, a manufacturing system of aerospace structural parts based on the Industrial Internet of Things is proposed. Key technologies such as general framework, technical requirements, integration model, interconnection and interoperability of IoT on equipment, technology and system are studied. On this basis, the five industry standards of "Research on the Industrial Internet of Things Reference Model for the Manufacturing of Aerospace Key Structural Parts", "Intelligent Marking of the

Manufacturing Process of Aerospace Key Structural Parts", "M2M for the Manufacturing Equipment Room of Aerospace Key Structural Parts", "Aerospace Key Structural Part Manufacturing Equipment and Processing", "Interconnection of Objects" and "Interconnection of Manufacturing Equipment and Manufacturing Systems for Key Aerospace Structural Components" are independently formulated. The standards establish an industrial IoT standard system suitable for the intelligent manufacturing of the key aerospace structural parts, which is shown in Fig. 6.5. An industry test and verification platform is developed, which is applied in three typical enterprises of manufacturing the aerospace key structural parts.

(1) Reference model for industrial IoT applications in aerospace key structural parts manufacturing

The production platforms of upstream and downstream enterprises involved are integrated and coordinated to focus on manufacturing the aerospace key structural

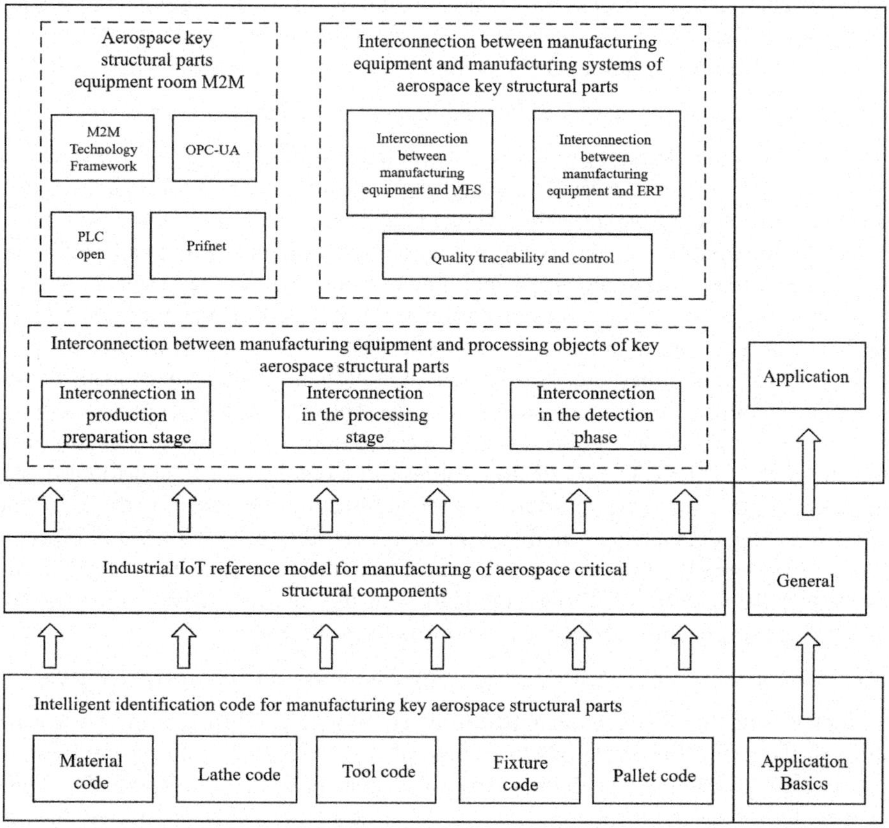

Fig. 6.5 Relationship of five interconnection standards

parts. The system has three levels of interconnection and system integration, mainly including:

(a) M2M based on OPC UA for manufacturing equipment room for key aerospace structural parts.
(b) Integration and interconnection of manufacturing equipment and processing objects based on intelligent IoT logos.
(c) Vertical integration of manufacturing equipment and ERP/PLM/MES manufacturing system.
(d) Application of Industrial IoT in the manufacturing the key aerospace structural parts.

(2) Intelligent IoT identification of manufacturing elements of the key aerospace structures

In order to realize the intelligent Internet of Things in modern aerospace, it is necessary to establish an intelligent identification (Tag) system to identify the production elements, such as, process information, manufacturing information and location information. Based on it, the marked objects can be controlled and managed in real time, and related information can be acquired, processed, transmitted and called.

(3) M2M standard for manufacturing equipment of the key aerospace structures

The production line integration is more urgent than other industries, which is related to the aerospace manufacturing industry involving process control, batch control, discrete machine control, and MES system integration, etc. Using OPC UA to interconnect equipment is an intelligent and automated method, which can better and faster realize real-time monitoring of equipment, achieve automatic control, improve production efficiency. The communication method based on machine-to-machine (M2M) can meet the interaction of real-time data, such as the interconnection of robots and assembly lines, etc. This bus-based interconnection method ensures reliability and stability of real-time data transmission. The used highly integrated network and communication interconnection can meet the needs of timely collection and transmission of data from various dimensions (horizontal, vertical, end-to-end), and meet the existing system and future system architecture. "Platform is the key", the data collection, data analysis, and system interoperability relying on the platform is the key to solving the information silos in the past. "Data is the core", "Data" flows and drives the Internet of Things application of the Internet of Everything on the basis of communication with each other through "platform".

(4) Interconnection between manufacturing equipment and processing objects

The interconnection between manufacturing equipment and processing objects runs through the entire life cycle of key structural parts manufacturing on the basis of intelligent logistics and distribution systems. The shape of the processing object will change, but its basic information and process information will be connected through intelligent logistics. Under the Internet of Things, in each link of the production workshop, intelligent logistics can accurately and punctually distribute the corresponding tools and materials.

Intelligent identification of IoT logos: The shape of each raw material and the various IoT logos printed on it are intelligently identified through the QR code identification system to accurately obtain the logo of the current raw material;

Assembly process matching: The assembly process is automatically matched through the identification of the Internet of Things. The involved elements, information, equipment are automatically found.

Assembly quality monitoring: Monitor the deformation and quality information of the assembled parts, and automatically judge the processing quality level.

IoT information integration: The data features obtained after assembly are automatically added to the raw material model information for future big data analysis, which is used for processing equipment monitoring and prediction, sheet metal processing quality traceability, and structural parts processing quality SPC.

(5) Interconnection between manufacturing equipment and manufacturing systems

The interconnection between equipment system and manufacturing system include connection and control. The connection can be explained from the three aspects. Firstly, processing equipment, logistics equipment, and storage equipment are connected using fieldbus. Secondly, the connection between equipment and control is a crucial part. The connection between the equipment layer and control layer is carried out in the form of industrial Ethernet. Finally, the transmission of data between the control layer and the big data platform is realized by industrial Ethernet.

Control is also important based on connections. The entire control layer mainly realizes the specific optimization of the lower layer (equipment layer) to improve the efficiency and safety of the operation of the lower layer; it responds to the commands of the upper layer (workshop layer), so as to realize the indirect control of the equipment layer by the workshop layer and even the enterprise layer. The first is that the big data platform obtains a large amount of data from the equipment layer. After intelligent analysis and decision-making of the data, the improved information obtained is transmitted to the industrial control system. The industrial control system then conducts automated and precise control on the underlying equipment. At the same time, the industrial control system will feedback the data generated in the control process to the big data center for further analysis and processing, so as to realize the control module of the intelligent workshop.

3. **Future aerospace intelligent manufacturing standard requirements**

The aerospace intelligent manufacturing standard has improved the overall level of digital design and manufacturing collaboration and integration. To a certain extent, it reflects the technical level and intelligent/digital factory construction experience of the advanced manufacturing industry. The resulting draft standard will help lay a standard foundation for its rise to aerospace industry standards, national standards, and even international standards.

At present, the relevant standards of intelligent manufacturing have played a significant role:

(1) Standardized process collaboration, process design and simulation, manufacturing process management, etc.;

(2) Open up vertical, horizontal, and end-to-end integration paths, and realize the smooth operation of data in the bus;

Aerospace intelligent manufacturing strives to build agile manufacturing capabilities and rapid conversion capabilities from research to batch production. Aerospace intelligent manufacturing vigorously promotes the intelligence of production site equipment and equipment networking and implements data collection and status monitoring throughout the production process. Aerospace intelligent manufacturing comprehensively promotes the construction of digital production units, digital workshops and digital factory.

6.2 Case Study of Intelligent Manufacturing Workshop for Typical Parts of Aero-Engine

Aero-engine is known as the flower of industry. It is the brightest pearl in the crown of modern industry. It has become an important symbol of a country's scientific and technological level, industrial base and comprehensive national strength. As the source of benefits for the entire enterprise, the parts manufacturing workshop is the basic link to ensure efficient, balanced and stable production of parts. To this end, this case will explain the successful implementation of intelligent manufacturing in the excellent manufacturing center of an aircraft engine Co., Ltd.

6.2.1 Aero-Engine Products and Their Production Characteristics

Aero-engine is a power device that provides power for the aircraft, which directly affects and determines the performance, safety, life, reliability and economy of the aircraft. As a highly complex precision power mechanical device, an aero-engine has tens of thousands of components integrated in a body whose size and weight are strictly limited. It is required to work reliably under conditions of high temperature, high pressure, high speed, and high load for a long time. In addition, aero-engines also need to meet special requirements in performance, applicability and environment.

The manufacturing characteristics of aero-engines are as follows:

(1) In order to maintain a competitive advantage in the fierce market competition, it is necessary to speed up the development of aero-engine models. Therefore, the production of aero-engine manufacturing is the multi-variety and small-batch production. Compared with large-scale production methods, the complexity and difficulty of running and management of aero-engine production enterprises have increased significantly.

(2) Aero-engine products have a large number of parts, which have complex supporting relationships. The requirements for complete sets of components and

products are extremely high, which ensure to deliver the aero-engine products on time.

(3) The production of complete aero-engines involves multi-unit and multi-department collaborative work. There is a close collaboration between each component manufacturer. The timely delivery of components from upstream production units has a greater impact on downstream production units.
(4) There are many types of aero-engine parts, including shafts, complex shells, casings, turbine disks and blades, etc. Their production process involves forging/casting and other wool production, roughing, processing, finishing, surface processing, physical and chemical treatment, shot peening and non-destructive testing and other production links. So the aero-engine can be producted collaboratively in multiple workshops, which also result in a long processing cycle.
(5) The production of aero-engine parts involves a large number of special tooling and special measuring tools. Controlling the processing progress of such special tools is the premise and foundation to ensure the production and delivery of parts on time.
(6) Most of the raw materials of aero-engine parts are precious metals such as superalloys, alloy steels and titanium alloys. Therefore, the raw materials and work in-process (WIP) of engine manufacturers generally occupy a large amount of funds. It is of great practical significance to improve the flow rate of WIP in production units.

6.2.2 Analysis of Receiver Products and Technological Characteristics

Viewed along the axis of the aero-engine, the receiver can be divided into front and rear ends. The front end is connected with other components of the compressor and assembled with various large-sized static turbine blades; the rear end is a complex flange structure, in addition to the complex hole system, there are also air holes distributed along the circumference.

The materials of receiver parts are mostly high-temperature alloys (GH4169, GH188, GH536, etc.), titanium alloys (T60, TC40, etc.) and other difficult-to-machine materials. Most of them are thin-walled annular parts, which have a cantilever structure and a split structure. In addition, receiver parts generally have many fine holes, especially in parts with high assembly accuracy such as mounting edges and flanges. The processing difficulties of such parts are mainly shown as follows.

(1) Deformation control of parts

Machining, welding and other process methods will cause different degrees of deformation. Therefore, the deformation of parts should be controlled by means of reasonably designed fixtures, reasonable arrangement of the processing sequence, and arrangement of a special flat datum operation before finishing.

(2) Measurement of precise dimensions

The diameter size, cusp point size, and feature point size with tolerance requirements within 0.1 mm are all difficult to measure. Geometric tolerances such as position and coaxiality can only be measured by a three-coordinate measuring instrument.

(3) Position guarantee of many groups of holes

Since each receiver part has multiple groups of fine holes and a large number of hole groups, and there is a complex angular relationship between each group of holes, any processing factors such as clamping and alignment during processing will lead to the deviation of hole positions. In order to eliminate various factors affecting the hole position to the greatest extent, a five-coordinate machining center must be used as far as possible to realize one-time clamping, one-time alignment and one-time measurement of the part hole group.

(4) Welding dissimilar alloys

When an assembly consists of welded parts of two different materials, it presents a huge challenge to the welding process. It is necessary to select reasonable radial and end face positions, and determine reasonable welding parameters and welding methods according to the force conditions.

6.2.3 Management Status and Existing Problems of Receiver Workshop

Process design and on-site production are the two main businesses of a workshop. With the increasing scientific research and production jobs, the original technical preparation and on-site production management of the receiver workshop have exposed the following problems:

1. **The main problems in the technical preparation stage**

   (1) Process design

   ① The part design information is required to re-input for the process model due to the lack of the 2D and 3D electronic drawings from the upstream design institute, which leads to a long production preparation period;

   ② The effective management of the typical processes from the software of process design is not performed, which leads that there is no feasible way to reuse the accumulated process knowledge;

   ③ The process resource information of processing equipment and typical processes in the workshop lacks effective management;

   ④ The quality and efficiency of information transmission are very low in those activities, such as product structure technical review, process plan design and process route design;

⑤ Tools application of and orders management are not done by the software, which results in lag in tooling applications, delays in production cycles, and repeated issuance of dispatch orders to tool factories.

(2) Tooling design
Generally, is based on two-dimensional model.
The three-dimensional design technology has not been fully promoted and applied as a replacement of two-dimensional design in tooling design. Design of the process and tools belongs to the serial design mode. The tooling design will be started after the process is completed and signed. The production preparation period is 3–6 months. It is a long time for product design.

(3) NC programming
There is no effective integration of CAD, CAM, and CAPP software; the management of NC code is chaotic; NC machining simulation technology is only verified and applied in scientific research projects and key parts processing, and has not been formally incorporated into the process design.

(4) Technical data management
The technical data management information is not exhaustive, and there is a lack of early warning mechanism. Search and traceability are limited. There are hidden dangers of data loss and leakage.

2. **The main problems in the manufacturing execution stage**

(1) Plan management lacks accurate empirical data support, which results in low plan execution. Production scheduling cannot grasp the overall situation.
(2) The planning management lacks certain flexibility. The planning cannot respond quickly for the unexpected situations in the production process.
(3) Due to the poor balance of planning management, material management is passive and chaotic.
(4) The tool management is relatively extensive. The relationship between the use, consumption, verification and planning of the tools is not carefully managed. Production consumption is ensured by high inventory of tools, which results in a large amount of funds.
(5) The collection of quality data at the production site is mainly manual collection. The information collection is more random. Quality control takes post-mortem inspection as the main means, and pre-prevention and in-process control are less powerful.
(6) Due to the lack of scientific and effective means of production status monitoring, accurate tracking and recording of standards such as spindle speed, production preparation, start-up time, and completion time cannot be carried out, which affects the accuracy of planned labor.
(7) Production costs control is lacked. Aiming to deliver on time as the production goal, production is often carried out regardless of cost for on-time delivery, resulting in unbalanced equipment load, overtime and high production costs.

(8) The information of various departments on the workshop is partial and scattered. The situation of "rich data and poor information" has initially appeared for decision-making.

The changes from the two aspects of production organization and information support technology are made to solve the above problems existing in the receiver workshop. In terms of improving the production organization, the new workshop organization mode of the excellent manufacturing center is introduced; in the aspect of information technology support technology, the intelligent manufacturing project characterized by information and digitization is implemented.

6.2.4 Analysis of Receiver Products and Their Technological Characteristics

The Center of Excellence (COE) is a brand-new workshop production organization method. The multi-product and multi-model production lines in the enterprise are classified according to the parts and components. A relatively independent manufacturing unit forms with the coordination of the departments of technology, production, craftsmanship and quality in the enterprise. The COE is responsible for the entire life cycle of the unit and has all the functions required for process design, procurement, manufacturing, testing and delivery.

1. **The internal organization structure of the receiver COE**

The organization of the receiver COE is shown in Fig. 6.6. The entire COE includes production department, technology department, quality inspection department and comprehensive department. In addition, according to the production line division principle of "specialization and small flow", the traditional production section is divided into 7 specialized flexible manufacturing units suitable for the overall structure of the receiver. Each manufacturing unit is responsible for a certain type of parts in the receiver products. After each unit completes its own task of processing the parts, it will be assembled into finished receiver and delivered to the final assembly workshop for complete aero-engine assembly.

2. **Internal and external business logic relationship of receiver COE**

Figure 6.7 describes the business division among the departments within the receiver COE, as well as the business relationship between the COE center and other relevant functional departments and workshops It can be seen that the entire COE center is similar to an independent, specialized small factory, covering multiple business functions such as planning and scheduling, process design, quality inspection, manufacturing units, material/tool supply, and financial management.

During the daily operation of the COE, the production planning and scheduling room is in the core and leading position. It is responsible for receiving the Material

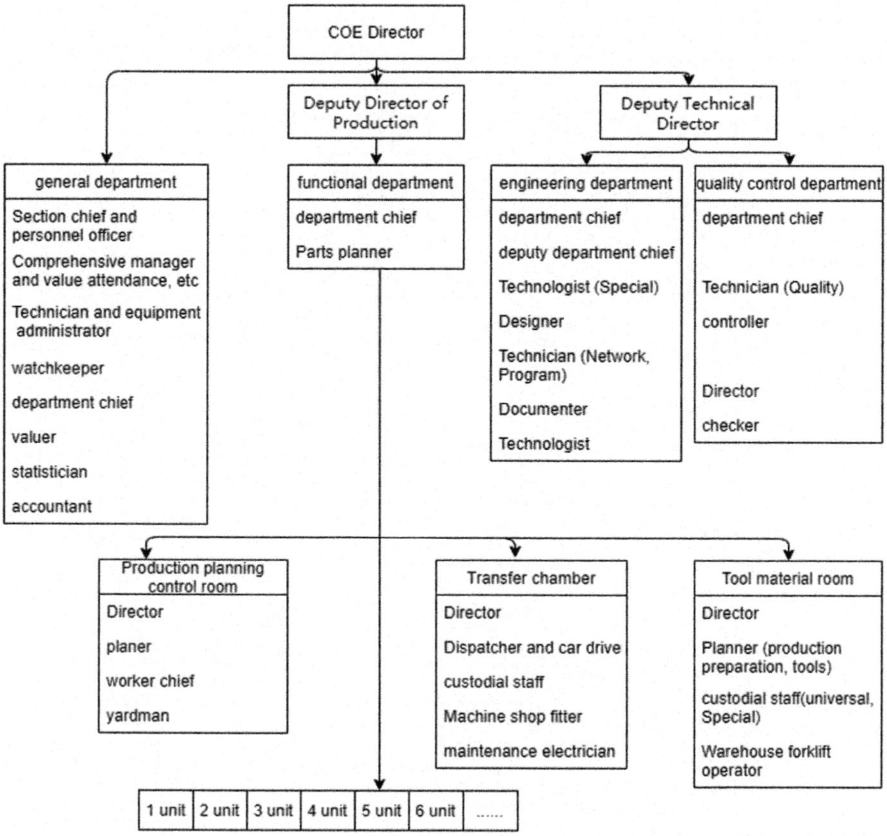

Fig. 6.6 Organizational structure of the receiver COE

Requirement Planning (MRP) order tasks issued by the superior production department, and arranging the monthly, weekly, Daily operation planning and scheduling, which promote the production preparation of wool, tools, equipment, technical data and other departments within the entire COE. The production planning and scheduling room is also the core and hub to formulate, arrange and coordinate the production tasks, shift assignments, processing progress and problems of each manufacturing unit.

6.2.5 The Main Content of the Implementation of Intelligent Manufacturing in the Casing COE

In the implementation of the intelligent manufacturing project, the receiver COE focuses on digitization and information. It takes process design and manufacturing

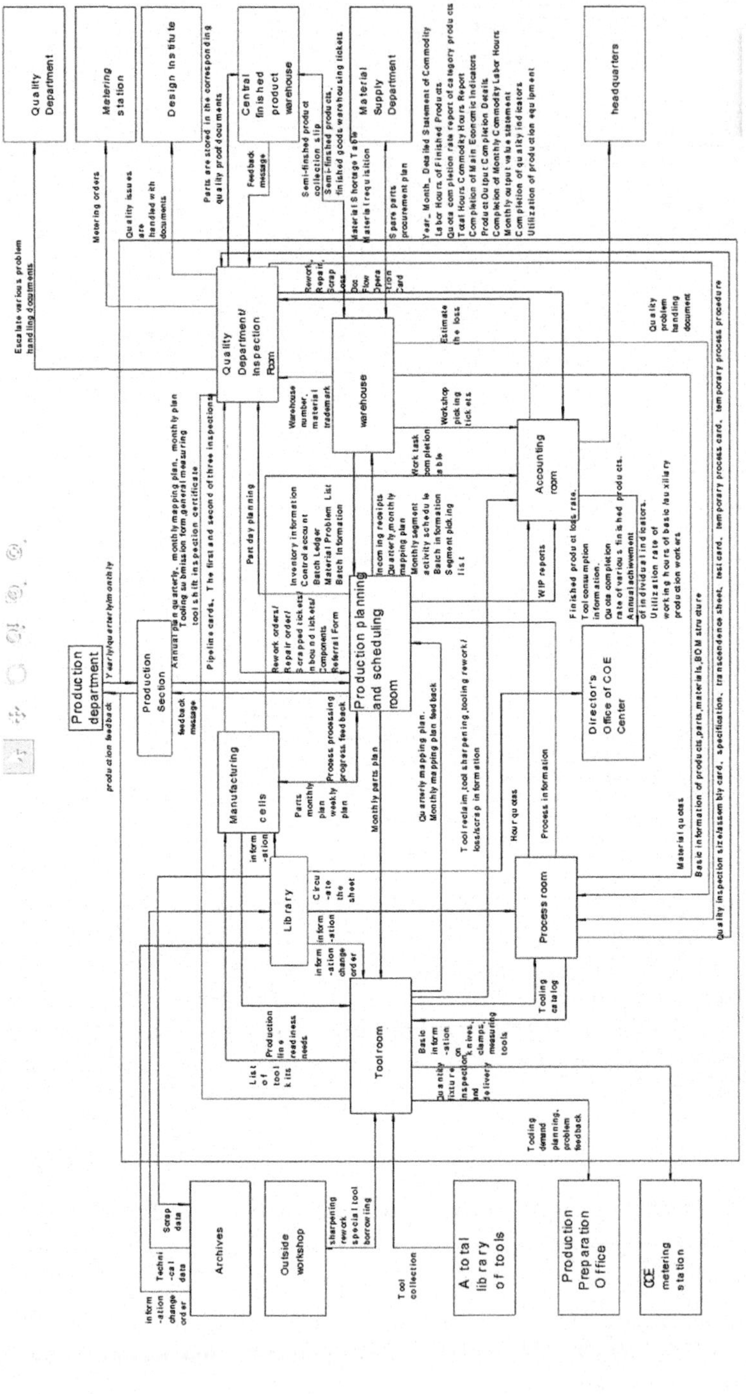

Fig. 6.7 Internal and external business relationship diagram of receiver COE

execution as the two-wheel drive, and takes the digital production line as the foothold, in order to realize the integrated management of "material flow", "information flow", "control flow", and "capital flow" in the production. With the characteristics of fast dynamic response and flexible manufacturing, it can meet the needs of different product types and manufacturing stages.

In the specific implementation of the intelligent manufacturing project, the receiver COE focuses on the three stages of technical preparation, production process simulation and manufacturing execution shown in Fig. 6.8.

The front-end of parts manufacturing includes process design, tooling design, NC programming, and cutting simulation. The digitization and parallelization of the technical preparation stage are realized with the support of the technical preparation application system. The process design method in the technical preparation stage is improved and optimized through digital means such as production line process simulation, processing route simulation and optimization. In the component manufacturing execution stage, the optimal allocation and efficient utilization of manufacturing resources such as human, machine, materials and methods are realized through the reasonable production planning and material tooling preparation. The strict control of the production quality is realized through real-time monitoring of equipment operation status and production progress. The main implementation contents of the three stages of technical preparation, production process simulation, and manufacturing execution are described in detail below.

1. **The content of the technical preparation stage**

As the front-end link of parts manufacturing, the implementation content of the technical preparation stage of the receiver COE is shown in Fig. 6.9. It mainly includes four subsystems: digital process design, digital tool design, NC programming and simulation, and quality analysis of trial-cut parts. Each system completes the technical preparations through product data management (PDM)-based CAD/CAM/CAPP/CAFD tool integration and information sharing.

As, the digital process design system is an important part of technical preparation system for the receiver COE. Its function can be described as following. The design of the process specification, process approval process and task management, process resource management, system management. The digital process equipment design system undertakes the design and management of process equipment used to process of parts and components, such as tools, fixtures, measuring tools, molds, various auxiliary tools, etc.

By integrating the functions of the CAD/CAM software system, the multi-axis CNC programming of complex parts based on the main process model is realized by calculating the tool position trajectory and obtaining the tool axis vector. The CNC program simulation is carried out to verify the NC code effectiveness and correctness. The cutting parameters are optimized based on the cutting parameter library. The technical status management of NC programs is realized through the PDM system. An integrated interface between the PDM system and the DNC system is built to realize the management and transmission of NC programs.

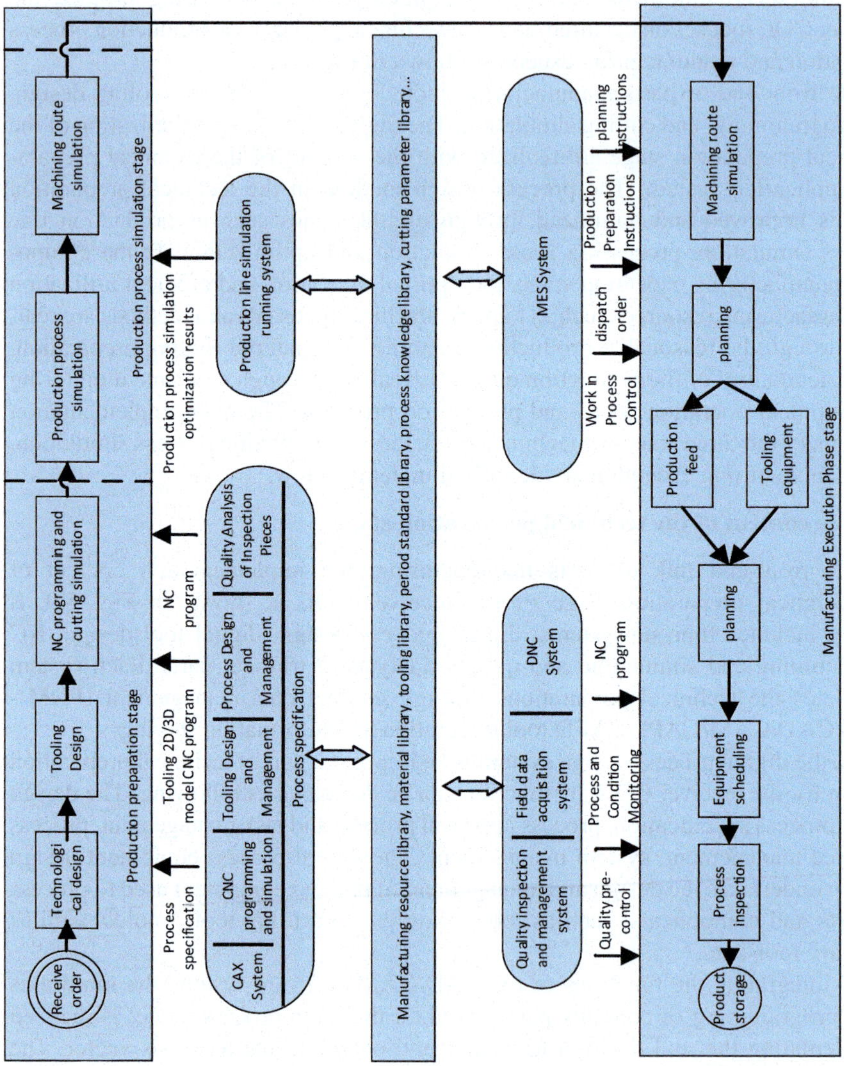

Fig. 6.8 Three main stages and contents of intelligent manufacturing in the receiver COE

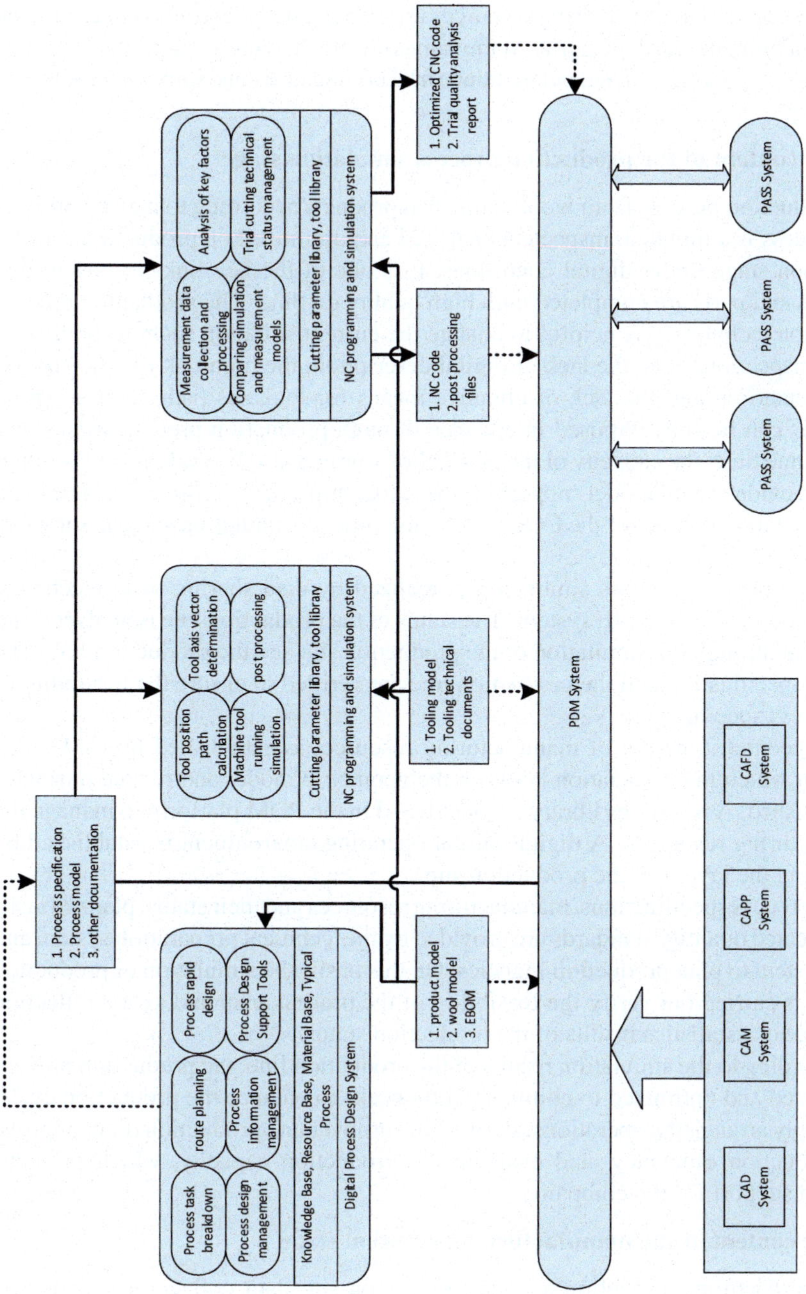

Fig. 6.9 The contents of technical preparation stage

The process of key operations is analyzed through the quality analysis system before the parts are officially put into production. The manufacturing operations, NC programs and manufacturing resources are evaluated and corrected to ensure the stable manufacturing of the parts. At the same time, the technical status management of trial cutting parts is carried out to reduce trial cutting costs and improve production efficiency.

2. **The content of the production process simulation stage**

The production process from wool to finished products according to a given production process is simulated, inspected, analyzed and optimized on production process simulation stage under digital conditions. It is one of the key links to ensure that products and parts are completed with high quality on time. The production process simulation technology is helpful to change the current situation such as the lack of digital inspection tools, the lack of digital description, the high risk of new process implementation, and the lack of effective means for logistics path control. Those problems can be fully exposed in advance through production process simulation. At the same time, the capacity of the production line can also be evaluated by simulation to provide digital model support for the leadership's decision-making. The main implementation content of the COE production process simulation stage is shown in Fig. 6.10.

The production process simulation is realized by data sharing of the technical preparation system and ME system. The status of the production line is analyzed and optimized through the simulation of the production site and the production line. The process specification and the production line are verified to ensure the feasibility of the process specification.

The geometric model of manufacturing resources is established by CAD software, on which the association between the geometric model and related attributes is established. A resources library is constructed in the PDM platform to manage the manufacturing resources. A digital virtual operating environment is established by simulating the layout of the production site.

The process specifications, manufacturing resources and their status, planned tasks and expected quantity standards are provided by the technical preparation system and MES system to plan production logistics. On this basis, the simulation of production process is carried out verify the feasibility of the process plan and give a reflection of the various statistics results of the production status.

According to the simulation results of the production line, the production process is analyzed and optimized to eliminate bottlenecks to balance the production cycle, reasonably arrange the operations, shorten the turnover and production time, improve the production efficiency, and evaluate the production capacity, which provides decision support for the company.

3. **The content of the manufacturing execution stage**

The manufacturing execution system (MES), on-site data collection and quality management system are implemented in the manufacturing execution stage of the receiver COE, as shown in Fig. 6.11. Among them, the manufacturing execution

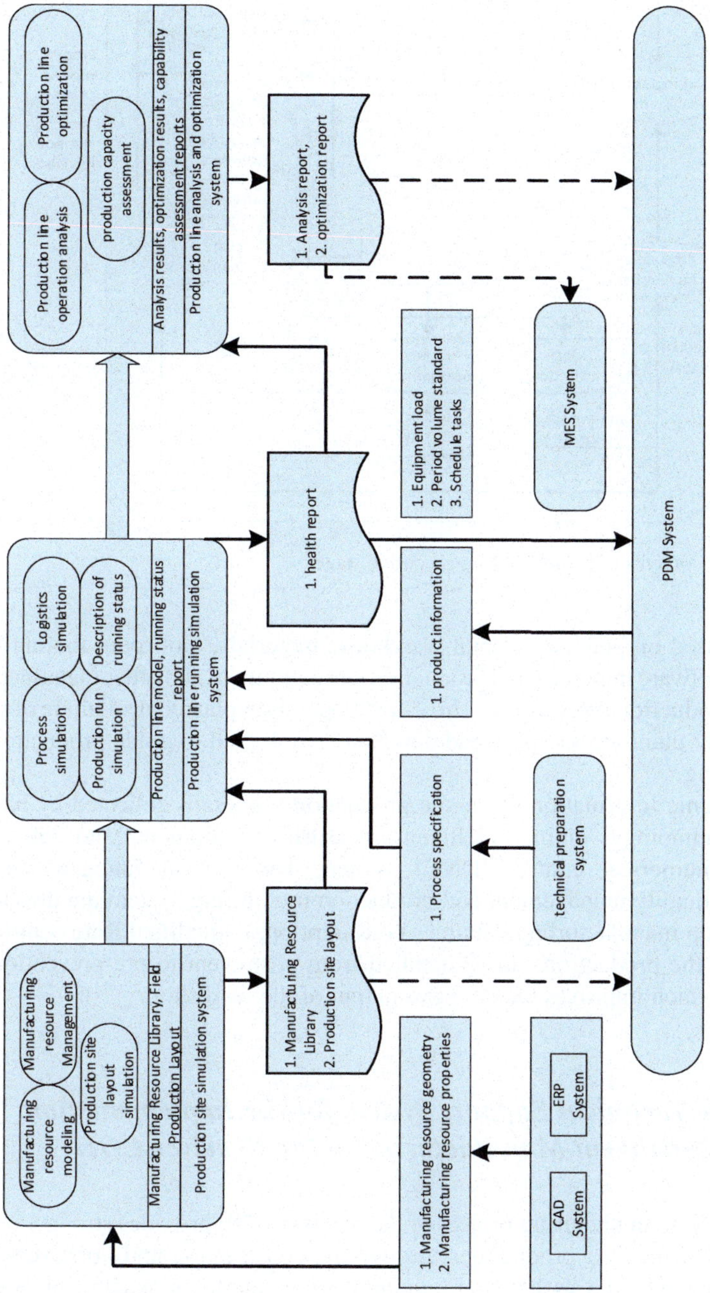

Fig. 6.10 The contents of production process simulation stage

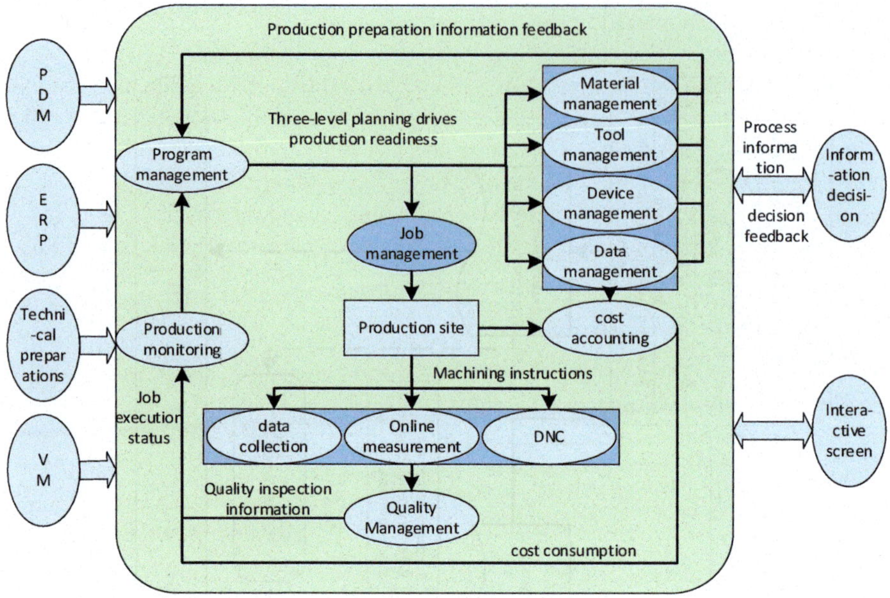

Fig. 6.11 The contents of manufacturing execution stage

system is based on planning to pull warehouse materials, tool room tooling tools, data room software materials and production site equipment and other manufacturing resources production preparation. Three-level operation plan control mode of workshop monthly plan, weekly plan and day plan are adopted to guide production site job scheduling.

The dynamic information of on-site production process is collected by barcode scanning technology, online intelligent measuring equipment, and integrating distributed numerical control (DNC) system. The control functions of cost accounting, quality management and production monitoring system are developed. The workshop manufacturing resources consumption, unqualified Rate control and problems in the production can be handled from post-event to pre-prevention and in-process, which improves the effective output of the workshop.

6.2.6 The Technical Support System for the Implementation of Intelligent Manufacturing in the Receiver COE

The intelligent manufacturing project of the receiver COE provides strong technical support for the receiver productions through the construction and improvement of important content such as hardware equipment, standards, management systems, corporate culture, application systems and digital tools. It has also laid a solid foundation for the promotion and implementation of intelligent manufacturing projects in key parts such as aero-engine blades, disc shafts, and disc rings. Figure 6.12 shows the technical support system of intelligent manufacturing of the receiver COE.

(1) Hardware support layer

A digital, networked and visualized production line hardware support platform consists of machine tools, flexible manufacturing units, automated three-dimensional warehouses, data acquisition, online measurement, visual display, and basic network is constructed.

(2) Standard specification layer

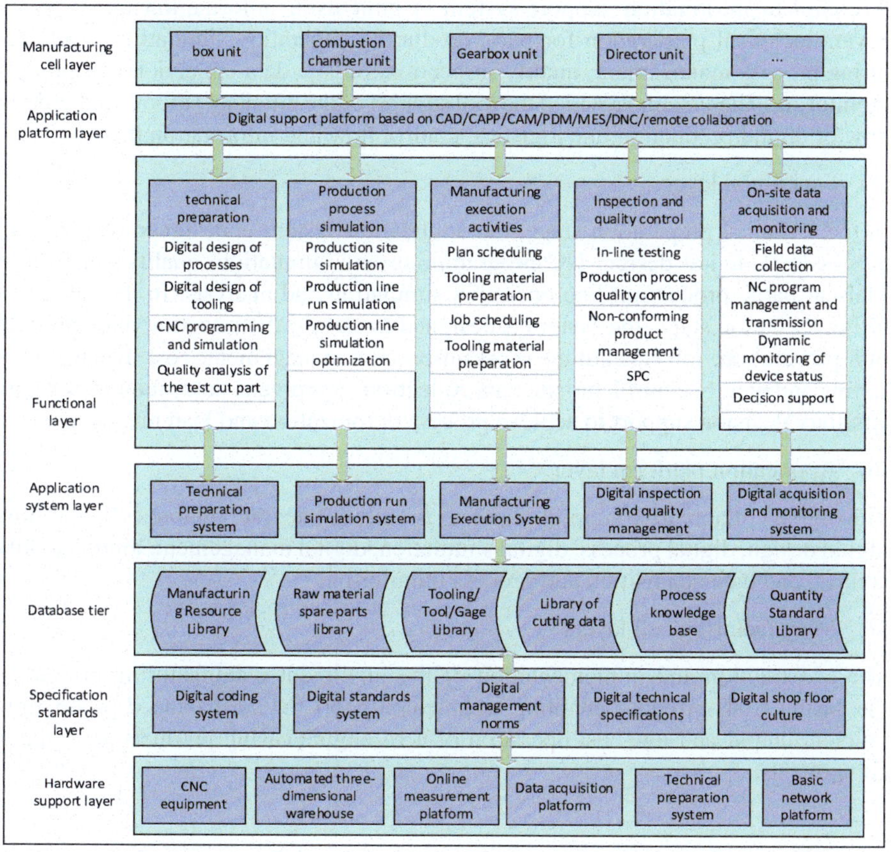

Fig. 6.12 The technical support system of the intelligent manufacturing project of the receiver COE

Focusing on the standard, system and cultural construction of the workshop, comprehensive construction is carried out from the aspects of coding system, process standard, inspection standard, management standard, system standard and workshop culture to ensure the standardized operation of the digital production line from the system and mechanism.

(3) Database layer

Focusing on the process parameters and quality standards of manufacturing resource required for parts manufacturing, a support database from processing equipment, raw materials, spare parts, tooling, cutting parameters, and quality standards is built to ensure the normal operation of the digital production line.

(4) Application system layer

A complete information support system is built using information tools, which covers technical preparation for parts production, operation simulation, manufacturing process management, quality inspection, on-site data collection and process monitoring. The system provided the tool support of the efficient, smooth and orderly flow of digital production line logistics, control flow and information flow.

(5) Functional layer

In the technical preparation stage, the application system is designed to carry out process/tool digital design, NC programming and simulation, quality analysis of trial-cut parts, production line operation simulation and analysis. In the manufacturing execution stage, the system is designed to carry out production plan, production preparation, job scheduling, work-in-process management, cost management, testing and quality control, on-site data collection, equipment status monitoring, and auxiliary decision support to achieve process optimization and lean management.

(6) Application platform layer

The digital collaborative support platform is built to integrate remote collaboration, digital design, digital process, digital simulation, digital management, digital quality control, and data acquisition and process monitoring.

(7) Manufacturing cell layers

The construction and implementation of the intelligent manufacturing project of the casing COE provides enabling technical support for the balanced production, efficient output and low-cost operation of aero-engine casing products and related components.

6.2.7 Design and Implementation of Receiver MES Software

Intelligent manufacturing is a complex and huge system engineering. Intelligent management and control in the production of parts and components is an important entry point for the implementation of intelligent manufacturing projects. The main body of intelligent production management and control is Manufacturing Execution System (MES).

MES can optimize the management of the entire production process from order placement to product completion through information transfer. MES can respond and process the real-time events occurring in the factory with current and accurate data to reduce non-value-added activities and guide the production operation process of the factory. MES can not only improve the timely delivery capacity of the factory, the circulation performance of logistics, but also improve the return on production. MES also provides mission-critical information about product behavior within the enterprise and throughout the product supply chain through two-way direct communication.

Based on the Workshop Manager 2.0 manufacturing execution system, combined with the comprehensive analysis of the organization, management status, business process, and process characteristics of the receiver COE of, MES software for the receiver COE was successfully implemented. In the process of designing, the overall operation control concept of the production line is adopted with planning as the core, process as the drive and performance as the foundation. The parallel preparation of personnel and technical materials realizes the efficient and coordinated operation of "man, machine, material, method and environment" in the entire workshop.

1. **Receiver MES operation process driven by monthly/week/day three-level process operation plan scheduling**

Figure 6.13 shows the operation flow of the receiver MES software driven by a complete monthly/week/day three-level operation plan. It can be seen that the receiver MES includes a complete information solution from receiving the MRP planning tasks of the upper-level production department to the inspection and storage of the finished receiver. Among them, the main line of planning and scheduling is taken as the leader and driver of the entire COE operation. The three-level control methods of the workshop-level monthly operation plan, the manufacturing unit-level weekly operation plan, and the team-level daily operation plan are adopted to decompose and progressive refine the receiver order tasks by level.

(1) The workshop-level monthly operation plan

The planner of the receiver COE center is specifically responsible for the preparation and optimization of the workshop-level monthly operation plan. The planning tasks are derived from the coarse-grained order task of the finished receiver. The planner adjusts and revises the order task according to the COE's internal inventory account, WIP processing progress, material input, etc. Then, he decomposes the BOM of the assembly, decomposes the parts according to the process route. Finally, he forms the official monthly production planning task through multiple simulation and pre-assessment, which guides the entire COE of the receiver. The production planning is send to the relevant departments and manufacturing units of the receiver COE to carry out the production of raw materials and spare parts.

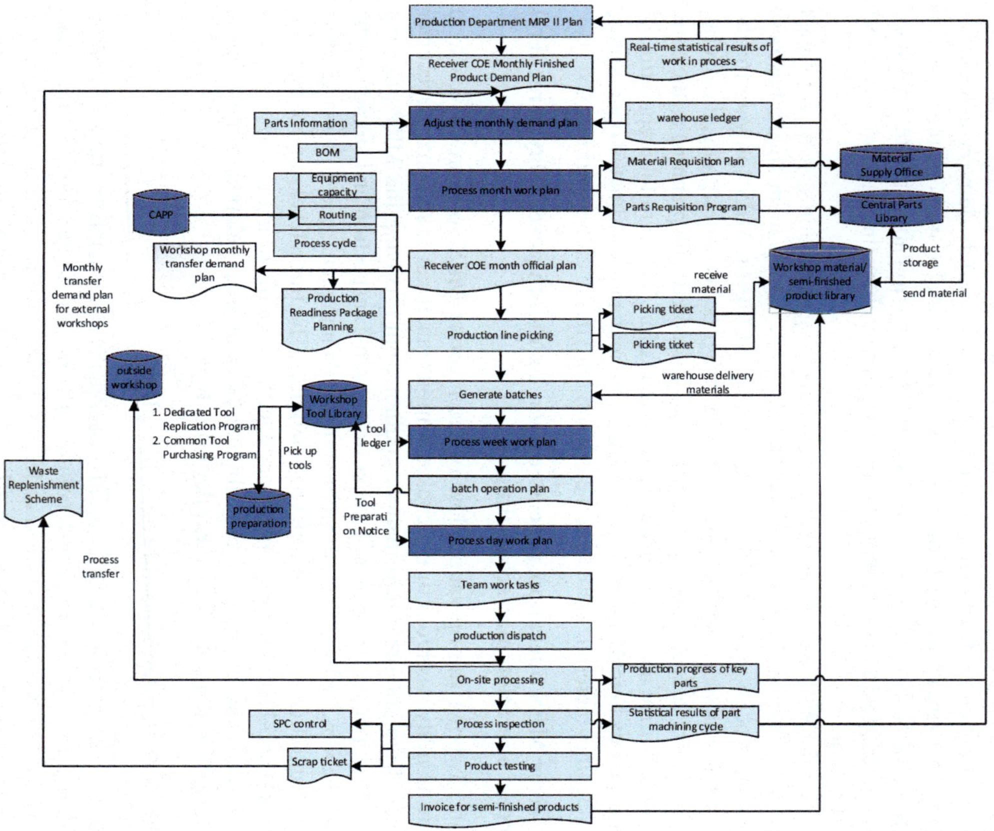

Fig. 6.13 The operation flow of the casing MES driven by the three-level operation plan and scheduling of the month/week/day

(2) The manufacturing unit-level weekly operation plan

After the monthly production planning task issued by the central planner is received, the actual picking operation is completed. According to the work-in-process progress, processing capacity, tool ledger in its manufacturing unit, the central planner working in manufacturing unit prepare the manufacturing unit-level weekly operation plan considering the time node of delivery requirements in the planned task.

(3) The team-level daily operation plan

The team is the lowest unit in the receiver COE, and it is also the specific undertaker of the processing task. The team management and production dispatch are closer to the production site. Therefore, after receiving the weekly work plan tasks issued by the unit planner, each team will prepare a daily work plan based on the factors such as the attendance of personnel in the team and the progress of process processing. Gantt chart of team work tasks is constructed to carry out daily work activities such as production dispatch and process processing.

2. **System function and external information integration of receiver MES software**

The receiver MES software adopts the B/S (Browser Server) framework based on Microsoft's Net platform. Its functions cover all businesses of the COE director office, the department of planning and adjustment, and the department of material. The software provides an information-based solution for the entire production of the receiver COE from order task to product delivery.

Receiver MES is mainly composed of 14 subsystems including plan management, job scheduling, production monitoring, inventory management, quality management, tool management, equipment management, cost management, data management, decision support, interactive signage, worker portal, basic data and system management. It is shown in Fig. 6.14.

In MES, the one-stop production management involving part task reception, plan release, material feeding control, tooling preparation, process processing, work-in-process circulation, finished product storage and statistical analysis is realized with the index of the part number. The information interface between ERP, PDM and the system of production preparation, material supply, central finished product warehouse is integrated in MES, as shown in Fig. 6.15.

3. **Three-level plan pull production readiness mode for receiver MES**

The planning and scheduling of three different objects, different granularities and different periods are the main line and core of the entire receiver MES. Figure 6.16 shows the hierarchical division, time period and pulling objects of the monthly/week/day three-level job plan. Through these three-level planning and scheduling, the parallelization and rationalization of production preparations such as wool, spare parts, tools, fixtures, measuring tools, molds, equipment, personnel, and technical materials in the receiver COE are pulled. During the preparation of the monthly operation plan, the planner of the receiver COE center receives the

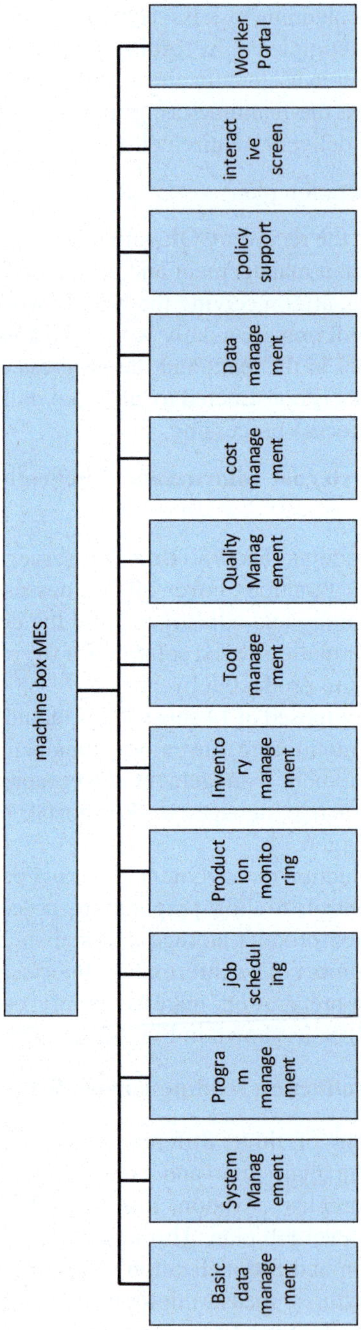

Fig. 6.14 System functions of the receiver MES software

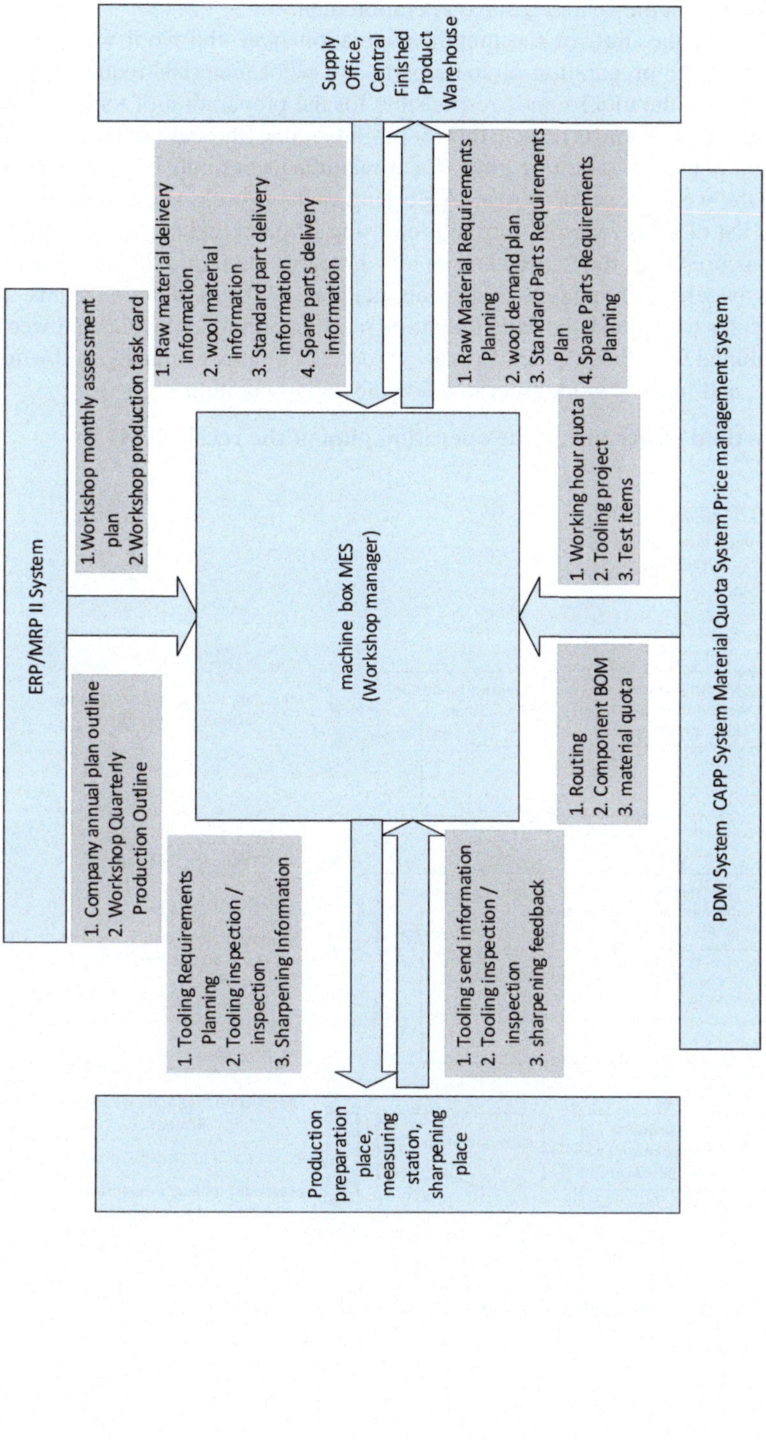

Fig. 6.15 Information integration of the casing COE and external systems

planned quantity and delivery node requirements in the MRP II task. Then, under the consideration of the quantity of work-in-process and semi-finished products, and their process route and process cycle, the planner formulate the monthly operation plan draft based on the coarse-grained operation plan.

According to the draft of the monthly operation plan, the wool warehouse is responsible for the preparation of spare parts and wool materials required for the production tasks; the tool room is responsible for the preparation of tools, fixtures, etc.; the outsourcing room is responsible for the outsourcing process and the advanced coordination and preparation of tasks. Each manufacturing unit is responsible for compiling the weekly operation plan. According to the process and technical documents, the list of tools required for the processing of parts and components for the next week is prepared; the material room also prepares the required parts and wool for the next week. Each manufacturing unit includes several production teams. The team scheduler formulates the daily operation plan according to the planned weekly tasks, combined with factors such as process route, equipment capacity, and worker attendance, and forms a work order to guide the daily operation.

4. **The method of preparing the operation plan of the receiver MES**

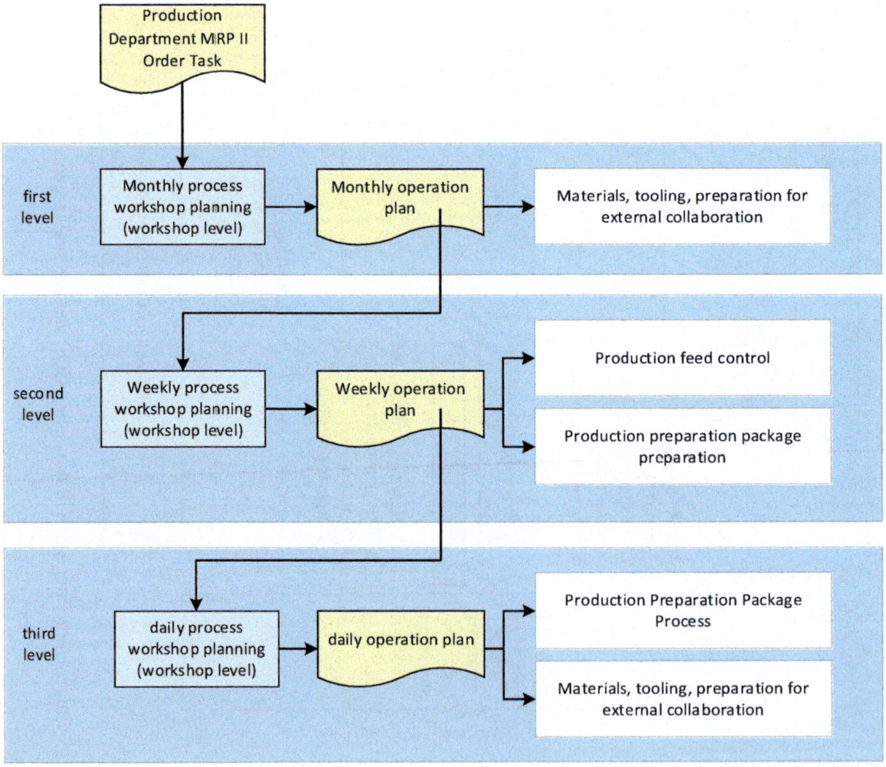

Fig. 6.16 Three-level planning and scheduling pull mode of receiver MES

In the receiver MES software, the monthly/week/day three-level process operation plan is the core and baton. Process planning (Scheduling, also known as job scheduling) is an execution plan in which the MRP II plan is decomposed and refined to specific operations. The plan is prepared according to the parts process route, process cycle, and in accordance with the progress of the work in progress and the actual production capacity. The plan specifies the time and quantity of the start and completion of each operation.

In fact, the operation plan is the most critical and specific end link to realize the company and the receiver center. The completion of daily and weekly tasks not only affects the completion of the COE monthly operation plan tasks, but also directly affects the smooth implementation of each part number plan of the MRP II system. Therefore, the company's products can only be delivered on time if the work section and manufacturing cell tasks are completed well.

At present, three methods: mathematical programming, intelligent optimization and sorting rules are used to solve the problem of process planning (or scheduling). Although the mathematical programming method can theoretically guarantee the optimal solution, it has high requirements for problem modeling and is only suitable for small-scale problems; the intelligent optimization method designs the feature model and neighborhood structure according to the characteristics of the job scheduling problem, and iteratively searches and intelligent optimization to obtain a satisfactory solution; the sorting rule is to prioritize the operation set according to a certain index, which can quickly obtain a feasible solution, but the optimization of the solution is poor.

The receiver COE is a typical multi-variety and small-batch production type. It has many production tasks, many product models, long process routes, long processing cycles, and frequent on-site exceptions. The two methods of mathematical programming and intelligent optimization are not suitable for the aero-engine production.

Figure 6.17 shows the processing logic of the receiver MES software when preparing the daily operation plan.

6.3 Exploration and Practice of Intelligent Manufacturing of Mengniu Dairy Products

6.3.1 The Situation of Mengniu Intelligent Manufacturing

"Internet +" and "Made in China 2025" are constantly promoting the development and reform of enterprises. In the context of the rapid internationalization of the industry, the market is changing rapidly, and the differences in consumer demand are intensifying, the demands for product safety are more stringent, the product development cycle is shortened. The precision, flexibility and intelligence of product services have put forward new requirements for enterprises. At the same time, cost

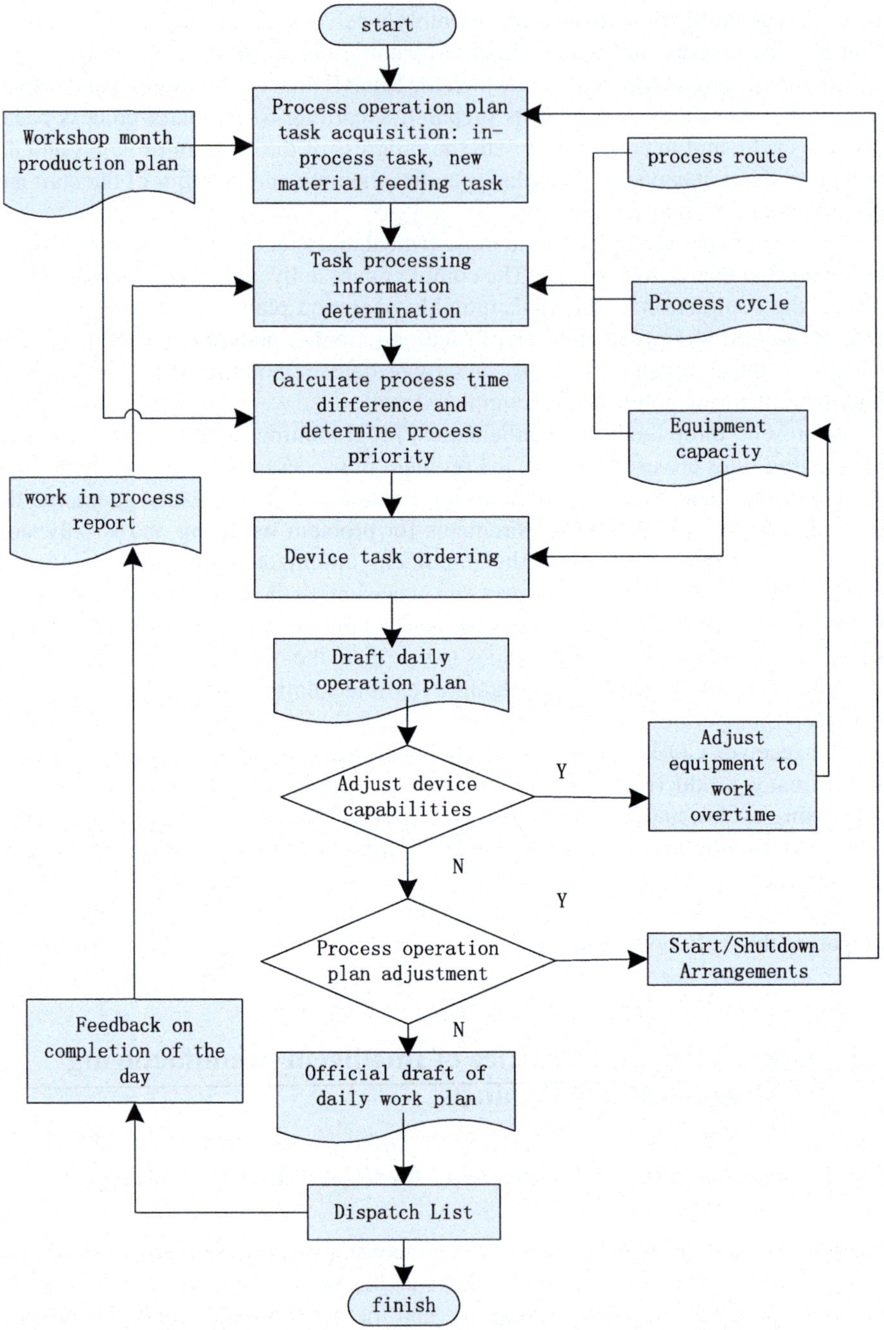

Fig. 6.17 The processing logic for the preparation of the daily operation plan

management has also put forward more stringent requirements for enterprises in the fierce industry competition. To solve the various problems currently faced, Mengniu Corporation would construct dairy intelligent factories based on current intelligent management foundation.

Mengniu intelligent manufacturing realizes the integration of four levels: at the management strategy level, applying BI for scientific analysis and decision-making; at the tactical level, applying ERP-SAP technology to manage the company's human, financial and other resources; at the executive level, applying MES Technology realizes digital management of production, quality, equipment, energy, logistics, etc.; at the acquisition level, the application of robots, smart instruments, PLC, sensors and other technologies to achieve automated production. The implementation of Mengniu's intelligent manufacturing breaks the information silos, realizes interconnection, and establishes intelligent energy management and control systems, automatic filling, automatic packaging, ERP systems, intelligent quality control systems. Mengniu's intelligent manufacturing system improves production efficiency, energy utilization, reduces operating costs, product defective rate and shortens product development cycle, and then promotes the overall construction of the industry, realizes the intelligent transformation of the traditional dairy industry to intelligence.

6.3.2 The Position of Mengniu Intelligent Manufacturing in the Intelligent Manufacturing System Architecture

As a complete process-based industry, Mengniu's intelligent manufacturing fully complies with all the characteristics of process-based manufacturing. The production process includes milk collection, milk purification, milk conditioning, pasteurization, ingredients, volume setting, UHT, filling, packaging, palletizing to warehousing, which is continual and fixed. In Mengniu's intelligent manufacturing, product orders are automatically analyzed, converted, and scheduled. Through the coordination and cooperation of various integrated and interoperable systems, materials, raw milk, quality inspection plans, energy supply, Warehousing location are provided to realize the real-time tracking of order task information, real-time monitoring of energy consumption, more efficient production, more reliable processes.

Mengniu's intelligent manufacturing basically covers all the content of China's intelligent manufacturing system architecture. Mengniu's intelligent manufacturing system covers design, production, logistics, etc. It integrates and connects PLM, MES, LIMS, EMS, WMS and other systems to realize the process interconnection of product design, production, traceability, energy management, logistics management, to also realize transparent production, one-click traceability, and data management. At the system level dimension, Mengniu's intelligent manufacturing covers equipment, control, workshops, enterprises, etc., and realizes the interconnection of people, equipment and systems, and achieves state awareness, real-time analysis, independent decision-making and precise execution. At the intelligent functions dimension, Mengniu intelligent manufacturing covers emerging business formats, resource elements, system integration, interconnection and information fusion, which has built the Mengniu intelligent factory cloud platform to realize the information fusion and interconnection of the whole process.

6.3.3 Introduction to Mengniu Intelligent Manufacturing System Architecture

As shown in Figs. 6.18 and 6.19, Mengniu's intelligent manufacturing system architecture consists of planning layer, execution layer and control layer. The planning layer is composed of enterprise resource planning (ERP), smart supply chain, product life cycle management PLM and other information management systems. The execution layer is composed of production execution system MES and its public function modules including production management, quality management, equipment management, cost management, report optimization and system integration. The control layer is composed of the automatic production equipment, which is used to receive, process, package and storage milk. The automation control equipment and information management system are integrated in-depth through the MES. ERP receives the production tasks of the smart supply chain, and sends the production plan to the production execution system MES. After automatic production scheduling and optimization, the MES issues production instructions to the production equipment at the control level. At the same time, the production task generates production methods and personnel instructions in the quality module, equipment management module, production management module, and cost management module of the MES. The equipment at the control layer executes production according to the MES production instructions, and feeds back the production status and production results to the MES in real time during the production execution. MES then returns the production results to the ERP. The production management granularity of the control layer, execution layer and planning layer is different. The control layer automatically collects production data in seconds and feeds it back to the MES. The MES analyzes the production status in real time in minutes, so as to accurately control the production progress and dynamically control production costs and product quality; ERP daily manages the completion of production tasks, costs, and quality.

6.3.4 Mengniu Intelligent Manufacturing Key Performance Indicators

Through the implementation of intelligent manufacturing, Mengniu has actually brought changes and benefits to the enterprise shown as follows:

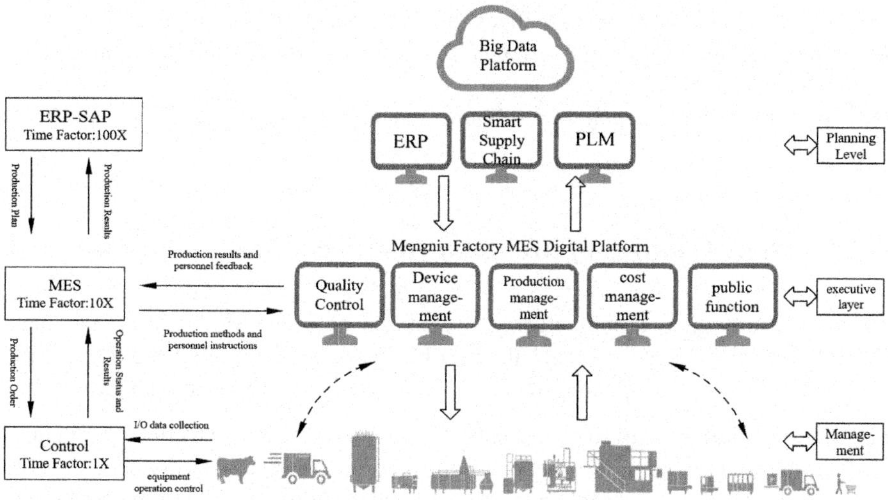

Fig. 6.18 Mengniu's Production System architecture

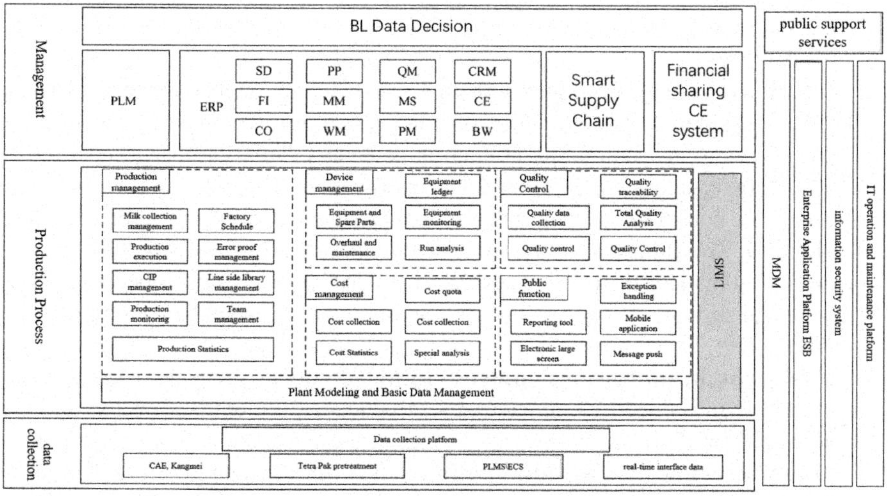

Fig. 6.19 Mengniu's intelligent manufacturing system architecture

(1) The speed of financial settlement is increased by 50%;
(2) Production efficiency increased by 20%;
(3) The whole process of digital automatic collection and monitoring can realize timely warning of abnormal data;
(4) Systematic and refined cost management and control, realize online monitoring and automatic accounting, and reduce operating costs by 20%;
(5) Systematic automatic production scheduling, quick response and meet the needs of market flexibility;

(6) The whole industry chain realizes one-key traceability of quality;
(7) The defective product rate is reduced by more than 20% by the automated quality inspection;
(8) The energy management is intelligent, and the energy utilization rate is increased by 10%.

6.3.5 Main Features of Mengniu Intelligent Manufacturing

Mengniu's intelligent manufacturing is centered on order life cycle management. The system analyzes and matches the best factory to produce according to the order based on big data. The specific plan is implemented from the following 5 parts:

(1) **Automation of quality control**: All inspection data is centrally managed through the LIMS (Laboratory Information Management System) system, which organically combines inspection and quality control to achieve automation of quality control, standardization of inspection processes, paperless inspection records, and automation of data collection. The full-chain information-based product control and traceability system is built by the application of the Internet of Things technology, LIMS system, and the terminal secondary traceability.
(2) **Intelligent energy management and control**: The energy management and control system uses intelligent instruments to improve the three-level energy measurement and automatic collection of energy data, wireless transmission, data storage, Internet technology. The energy data of various production plants across the country is summarized to display on the energy management center platform. The function of energy cost calculation, real-time monitoring, automatical statistical analysis of the factories across the is developed to provide technical support for energy-saving potential tapping and optimized production scheduling.
(3) **Intelligent product development**: PLM (Product Lifecycle Management) system is introduced to manage new product requirements, develop and test new product.
(4) **Manufacturing process automation**: The MES (Manufacturing Execution System) system is introduced to deeply integrate the automation and information of the manufacturing workshop to realize the intelligence and flexibility of the manufacturing process.
(5) **Logistics intelligence**: In the process of project construction, the three-dimensional warehouse automatic stacker, the loop-through trolley and system, and the key process equipment are all equipped with high degree of automation equipment, the intelligent function of equipment is realized using a variety of intelligent functional components and a unified standard communication protocol. Cooperated with WMS (Warehouse Management System) system to realize automatic data collection, some equipment realizes two-way interaction and control with WMS system. Vehicle position and product temperature

during product transportation are full-time monitored. The remote monitoring and management function of national warehouse accounting is realized relying on Internet of Things technology.

6.3.6 Mengniu Intelligent Manufacturing Implementation Steps

Before Mengniu implements intelligent manufacturing, a detailed investigation of the business needs of the production executive, management and decision-making layers is summarized according to the degree of impact on cost, quality and efficiency. The business scope of Mengniu's implementation of intelligent manufacturing is determined based on the refinement and analysis of Mengniu's excellent running mode.

After determining the business scope of intelligent manufacturing, Mengniu Intelligent Manufacturing will carry out automated upgrades at the control layer, including: milk collection, sterilization, filling, packaging, palletizing, inbound and outbound, logistics and market terminals. The automation upgrade and transformation of each part are included in Mengniu's intelligent manufacturing.

Milking adopts a fully automatic robotic milking system. Through automatic milking, the rancher can get rid of the heavy and trivial routine milking affairs. Firstly, the automation of the milk collection and sterilization process, in which all pumps and valves are controlled by PLC to realize the automatic flow of milk according to process control parameters, which achieve product traceability and traceability, and automatic recording of equipment operating parameters for easy traceability. Secondly, the automation of filling, packaging and palletizing, precise filling system and aseptic filling system automation are achieved by automatically detecting stable filling flow through flow sensors set on the filling equipment. Box, packing, sealing, palletizing automatically are realized using PLC control system and robot, which improves production efficiency and reduce labor intensity. Thirdly, the automation of in and out warehouse, multi-axis manipulator, AGV, RGV, aerial suspension, stacker, code line, conveyors, circular shuttles are integrated in automated three-dimensional warehouses to improve the speed and ability of products in and out of the warehouse. In terms of logistics automation, the GPS system is introduced to timely monitor important indicators such as vehicle location and temperature changes. In terms of market terminals, the product packaging box adopts QR code technology, and the product information can be directly obtained by scanning the code with smart tools such as mobile phone terminals.

While building the automation upgrade of the control layer of the production equipment, information construction is carried out in the planning layer (management layer). The information construction is mainly composed of PLM, ERP, MES, logistics management system and production automatic control system. The concepts, methods and technologies of ERP are applied to establish an integrated management platform based on standardized business processes and finance.

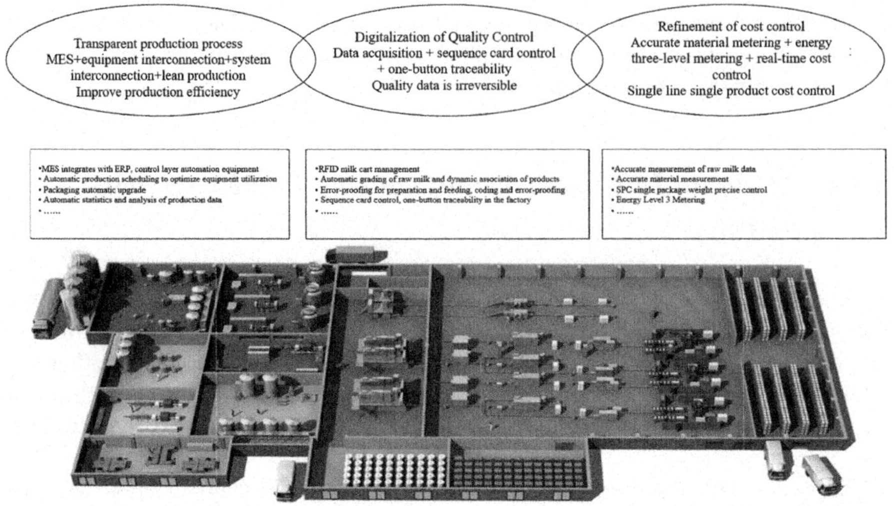

Fig. 6.20 Information construction of Mengniu intelligent manufacturing

The PLM product lifecycle management platform provides enterprises with reliable, open and flexible product design, collaboration, workflow and knowledge management functions.

MES realizes the collection, processing, storage, utilization and management of data. On this base, production management, weighing and ingredient management, quality control, material and warehouse management, equipment management and equipment OEE, production electronic batch records and production performance evaluation can be carried out.

By interconnecting advanced local automation and information islands in these production links, the market demand is transformed into the decision basis for supply planning, production execution, and logistics transportation through the planning department. The quality, equipment, cost and other elements are controlled and coordinated, and plan changes are handled, so that production can run efficiently, orderly, with low cost. The quality, efficiency, and cost-related indicators in the entire process are summarized and analyzed to provide decision support for management. Finally, intelligent manufacturing is achieved by seamless connection. The details of information construction is shown as Fig. 6.20.

6.3.7 Standardization Status and Demand of Intelligent Manufacturing of Dairy Products

Mengniu has summarized and refined good practices in the implementation of intelligent manufacturing to be three sets of intelligent manufacturing standards draft:

1. **Technical Requirements for Intelligent Energy Management and Control System**

    (1) Equipping rate and performance requirements of measuring instruments: The instrument has functions such as data reading, data memory, and data remote transmission;
    (2) The correctness and real-time nature of data collection;
    (3) Correctness of remote control: accuracy of data transmission and integrity of historical data.

2. **Operation Management Requirements for Intelligent Factory Workshop of Dairy Products**

The network and communication should be unblocked to meet the requirement of the intelligent dairy factory/workshop. The data collected by SCADA in the production process can be interconnected with the data of MES production orders, which can trace the data of the factory/workshop on demand. MES data interacts with ERP system data to provide more accurate decision-making basis.

3. **Information Specification for Dairy Products Laboratory**

Integrate the CNAS17025 laboratory management system into laboratory information management. The connection of inspection instruments, automatic upload of test data, electronic original records, automatic generation of inspection reports, real-time configuration of inspection methods, and automatic implementation of quality control plans are standardized. The technical guidance standard for the establishment, application and management of product laboratory information system is formulated to comprehensively solve the lack of laboratory testing automation and information application standards, improve the intelligent level of laboratory testing management and control.

As the most core and complex equipment in the production process of dairy products, dairy product manufacturing equipment plays a vital role in the production efficiency and quality of dairy products. As a fast-moving consumer goods with high requirements on quality and timeliness and huge market demand, the production process of dairy products is required to keep the continuous production and avoid to affect the freshness of the product. However, as the number and complexity of core filling equipment continues to escalate, traditional on-site equipment maintenance is increasingly difficult and maintenance costs continue to increase. In particular, dairy production lines are generally close to the place of consumption, resulting in production lines all over the country or even all over the world, causing great difficulties in the operation and maintenance of equipment. It is difficult to detect and locate equipment faults in time. The quality and production efficiency of products are difficult to improve, which brings many uncertain factors to the operation and development of enterprises. Therefore, it is urgent to develop remote operation and maintenance technology and related series of standards for dairy production equipment. Through the construction of a remote operation and maintenance service platform, the standardization of the remote operation and maintenance of the production and manufacturing equipment of dairy enterprises can be realized to ensure the safety and efficiency of dairy production. On the one hand, it is of great significance to improve the productivity and competitiveness of dairy enterprises. On the other hand, in the process of remote operation and maintenance of equipment, a large number of equipment operating parameters are collected to provide equipment manufacturers with more information for subsequent product improvement and design.

6.4 Gree's Air-Conditioning Smart Factory

6.4.1 Factory Network Layout

Changsha Gree HVAC Equipment Co., Ltd. ("Changsha Gree") continued to invest a lot of energy to implement the construction of the industrial network to meet the production requirement from planning. All key equipment interconnected in the park, which also connect people and data through the group network using professional factory management software. On the premise of ensuring information security, Changsha Gree achieve the goals of improving efficiency, reducing costs, and reducing resource use. Its network architecture has the following characteristics:

(1) **Comprehensiveness**: 5 workshops and 3 warehouses in the park have full coverage of wired and wireless networks without dead ends, providing a guarantee for equipment networking.
(2) **Efficiency**: The backbone of the system network is 10 Gigabit level, and the desktop network is Gigabit level.
(3) **Flexibility**: The system adopts standard transmission cables and related connection hardware, modular design, all channels are common, and all equipment can be opened and changed without changing wiring lines.
(4) **Reliability**: The system uses high-quality materials and combined crimping to form a set of high-standard information transmission channels. The application system use point-to-point termination to ensure the reliable operation of the entire system.
(5) **Security**: The internal network is physically isolated. The encryption is adopted as an active defense method to ensure the security of the internal network system.

6.4.2 Selection of Digital Equipment

In the early stage of equipment planning, Changsha Gree has widely selected digital and automated equipment. The equipment is perceptible and has certain data processing capabilities. It can collect dynamic operating data of the equipment in real time. It can independently choose the data analyzed and stored on a remote location according to the preset model. On the other hand, NC machines are selected to machine the parts based on feedback from their own senses or digital commands sent remotely.

Faced with the different PLC interfaces and protocols of equipment from different manufacturers, the communication interfaces, types and protocols were unified to be the standards agreed in advance, which laid the foundation for the interconnection of equipment.

6.4.3 Application of Various Identification Technologies

Through bar code and RFID information technology, the batch management of incoming raw materials, the quota picking of production orders, the production status of the production line, and the accuracy and systematization of data operations in the production process are realized. The tracking control of product is established to trace the production from raw materials to parts, components and productions, which completes the internal process traceability management, the quality data management and the source traceability external entry and exit.

(1) **Barcodes batch management of incoming material** The handheld terminal scans the barcode information through the wireless network to realize the data synchronization of storage and outbound, achieve the consistency of physical logistics and information flow. The person can query the quantity of goods, storage location, batch and other information through the handheld terminal to improve warehouse management quality and work efficiency.
(2) **The transparent management of the production process** The RFID tag is bound on the product through the MES system. The relationship between the production batch and the RFID tag are established to record and track the production process, which can trace to the potential problems of operations materials and models.
(3) **Refinement management of finished product storage location** Through automatic equipment stacking board scanning barcode identification, the workload of finished product warehousing, warehouse transfer, and delivery is reduced to ensure the accuracy and efficiency of finished product inventory information. The refined management of finished product storage location is realized through the binding of barcode identification and storage location. It can quickly and accurately realize product location positioning.

6.4.4 Deep Integration Application of ERP and MES System

In the current production management of Gree, ERP is the core of planning and resource management, and MES business structure is the core of the manufacturing site. Focusing on the integrated application of the two platforms, Gree has independently developed a data-driven manufacturing information collaborative management platform based on ERP, MES to realize the horizontal interconnection, data flow, conversion and mutual recognition between factory software. The data synchronization rate of the two systems reaches 100%.

The plan management of ERP and the field management of MES are deeply integrated through this platform shown in Fig. 6.21. The main functional modules of the platform are divided into the following three:

(1) The coordination of production planning and logistics is realized through a complete set of production scheduling management system;
(2) The coordination of distribution planning and line-side demand is realized through the electronic picking system (SAM);
(3) The coordination of the implementation process and accounting management is realized through the landing backflush system.

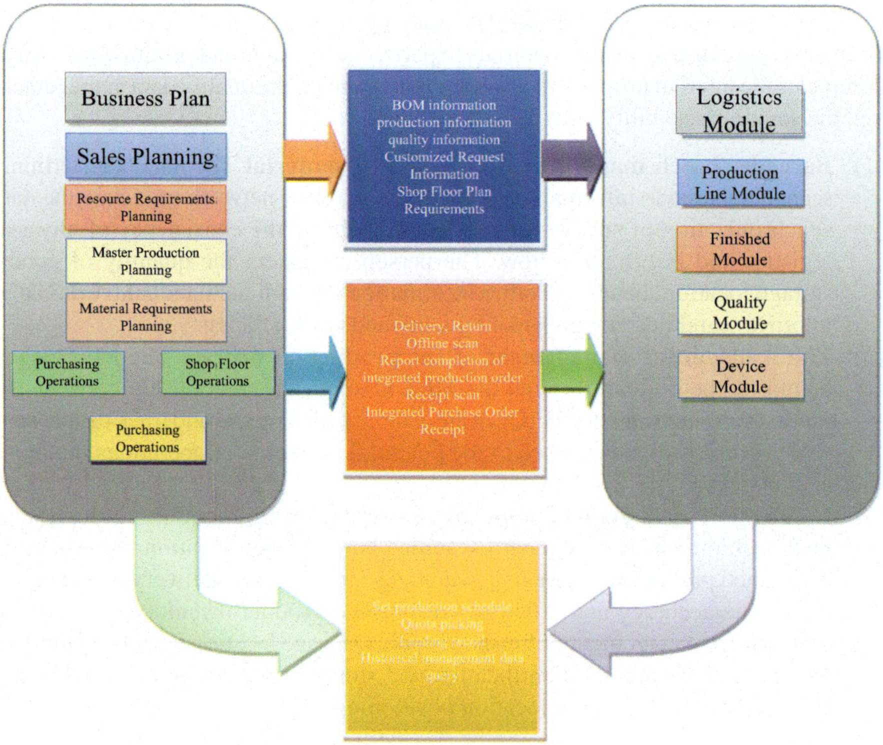

Fig. 6.21 Deep integration of ERP and MES systems

6.4.5 Post-implementation Effect of Industrial Network

Changsha Gree Factory builds the software architecture of the industrial network based on the SCADA system platform, as shown in Fig. 6.22. The bottom layer is the workshop equipment layer. The real-time running data of equipment is collected through the IGS interface software connected to the PLC of the equipment. The next layer is the SCADA core application layer. The related application business of production order management, production process monitoring and equipment call alarms is developed and displays through iFix software. On the one hand, it can receive the production orders arranged by MES from the MES layer. The completion of the execution of production orders will be fed back to the MES system. On the other hand, through the interface layer, the MES can also feed back some real-time data of the equipment production and operation required by the MES, such as equipment failure alarm and other information. The MES can make corresponding processing according to the real-time information of the production site.

Through programmable logic controller (PLC), data acquisition and monitoring and control system (SCADA) control, the 100% interconnection of the key equipment in each branch is achieved to real-time collect the information of equipment action

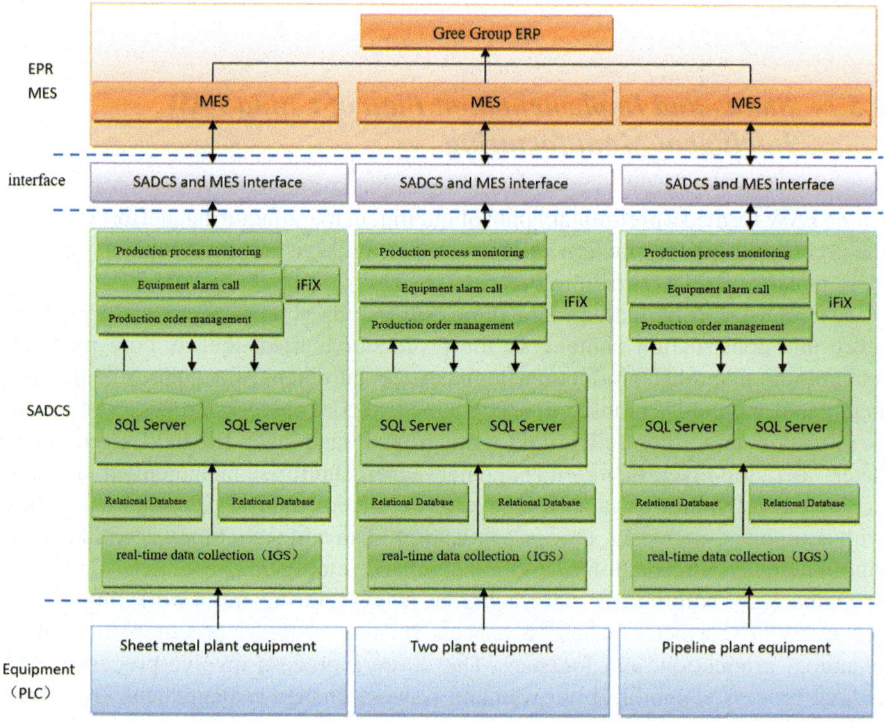

Fig. 6.22 Software architecture of industrial network

and status, which greatly reduces the manual collection of production process data, also reduces the company's operating costs. The main points of the formation are as follows:

(1) Realize 100% interconnection of key equipment in 5 major workshops;
(2) Real-time collection of equipment operating status;
(3) Effectively reduce the human resource investment in equipment management and data collection;
(4) Improve the response speed of equipment abnormality;
(5) Data mining and application in depth.

6.5 China FAW Intelligent Manufacturing Exploration

China F.A.W. (First Auto Works) Group Co., Ltd. (China FAW), as the eldest son of China's auto industry, has always adhered to the innovation-driven strategy in the course of more than 60 years of development, and has made outstanding contributions to the development of China's auto industry. China FAW takes intelligent manufacturing as a key project of technological innovation for the new breakthroughs, which provides the strong support for future transformation and development.

6.5.1 Status and Implementation Plan of China FAW Intelligent Manufacturing

China FAW formed a preliminary plan of intelligent manufacturing platform involved the process design, production and manufacturing, logistics management, and post-sales management, as shown in Fig. 6.23.

A framework is built around the three dimensions of professional field, coverage level, and construction content. In the professional field, the six core majors of technology, planning, production, logistics, procurement and quality in the large manufacturing field is focused to open up the two core business processes of product development and order delivery. In terms of coverage level, the field layer, control layer, operation layer, factory management layer, enterprise management layer, and ecological collaboration layer are covered. The construction of digital capabilities with equipment, networks, processes, systems, data and technology is realized from automation and information to digitalization and intelligence.

China FAW summarizes the intelligent manufacturing framework into 20 core processes in 6 aspects. These 6 aspects involve process, procurement, quality, planning, production, and logistics. The 20 core process involve process coordination, process capability improvement, process change management, outsourced parts procurement, supplier collaboration, outsourced parts quality closed-loop management, supplier quality improvement, and complete vehicle 20 core processes

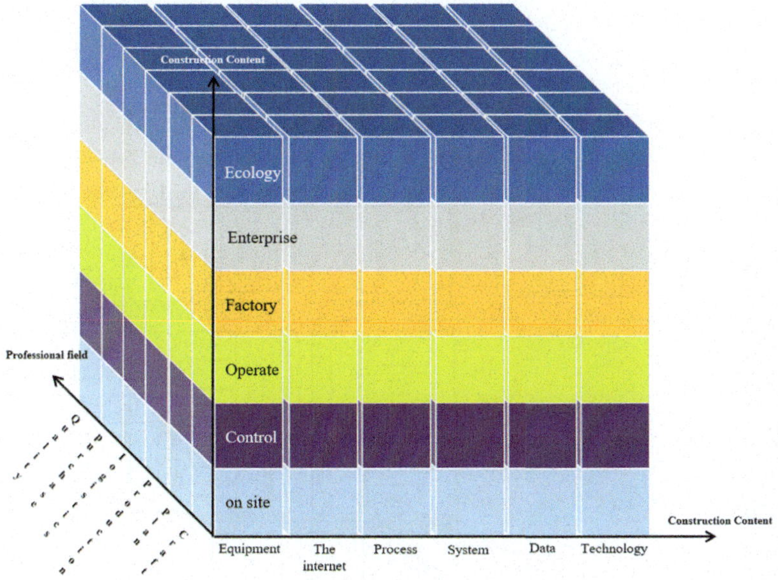

Fig. 6.23 The intelligent manufacturing framework of China FAW

including quality closed-loop management, testing resource management, production planning, material planning, production process, on-site management, personnel management, equipment management, safety management, energy management, logistics planning, on-site logistics, and on-site logistics. The core processes are shown in Fig. 6.24.

On this basis, the application and construction of innovative technologies will be carried out on 21 major projects. They are digital process platform construction, process virtual simulation construction, factory construction engineering 3D simulation construction, procurement management platform optimization, quality management platform construction, digital online monitoring technology application, flexible supply chain planning system construction, energy saving and emission reduction technology application, manufacturing Platform construction, equipment predictive maintenance technology application, information traceability technology application, digital production operation technology application, parts logistics system optimization, intelligent logistics technology application, factory digital central control platform construction, factory digital/intelligent equipment construction, digital factory network And IT infrastructure construction, security protection and system construction, digital factory data asset construction and governance, intelligent application (industrial Internet platform) construction, collaborative field application construction and transformation. Figure 6.25 shows the innovative technology application and construction projects of China FAW Intelligent Manufacturing.

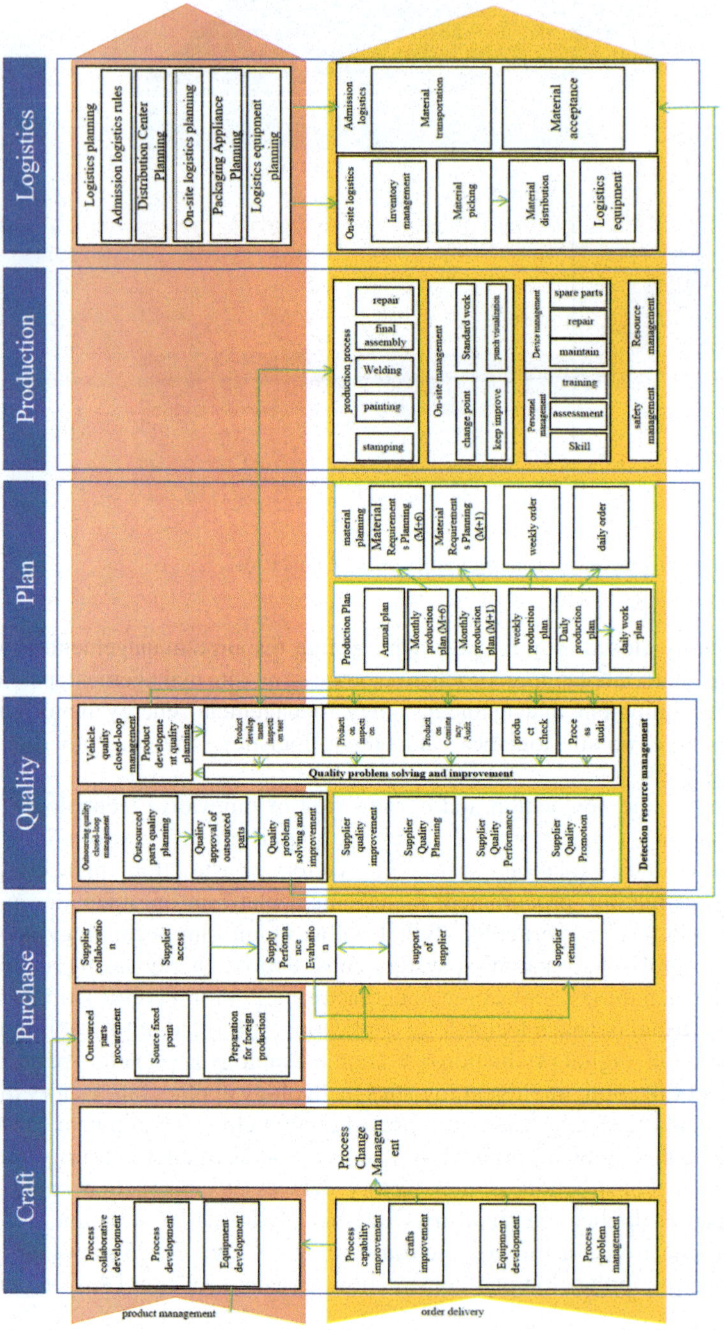

Fig. 6.24 The core process of China FAW intelligent manufacturing

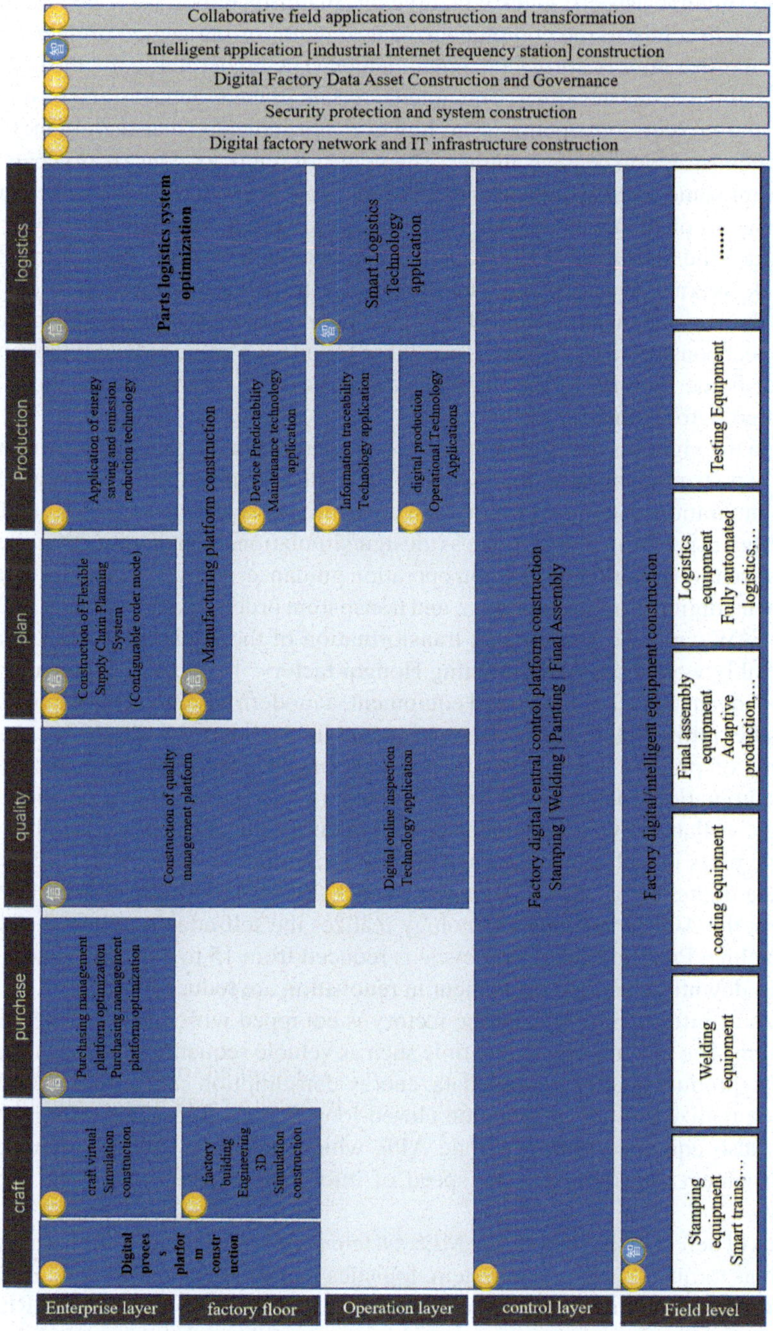

Fig. 6.25 The innovative technology applications and construction projects

In addition, the key principles of implementation of intelligent manufacturing construction are five aspects: quick win, urgency, maturity, importance, and foundation. Quick win refers to the comprehensive evaluation of the degree of value and the speed of effect after the project is implemented. The higher the score, the higher the value and the faster the effect. The urgency refers to the urgency of the business needs for the project construction. The higher score indicates that the urgency of the project construction needs is higher. The maturity refers to, construction can be prioritized in some areas where business management is relatively mature because of the small cost of change management. The management process can be optimized first to reduce the resistance to subsequent implementation in some areas with relatively weak internal management. The importance refers to the importance of the project to the relevant business field. The higher important value indicates the necessity of project construction is higher. The higher foundation value indicates the more the project is the premise of the implementation of other projects, and should be implemented before other projects.

Through the vigorous development and construction of intelligent manufacturing, China FAW is striving to realize the key scenes of digital factories, including: supplier online collaboration, intelligent production scheduling, intelligent logistics, intelligent quality inspection, digital process design/simulation, digital process collection and traceability, digital production operation guidance, equipment/energy smart management, digital monitoring center, and transparent orders, as shown in Fig. 6.26.

China FAW carried out intelligent transformation of the H platform and L platform assembly workshops of the existing Hongqi factory. Through applying a large number of advanced technologies and equipment, a modern assembly workshop that integrates flexibility, intelligence, automation and information is built.

In terms of process equipment, through applying advanced technologies such as robot automatic transfer, robot automatic assembly, optical camera positioning technology, collaborative robot, automatic tightening, the automatic assembly of many large parts is realized to improve the assembly quality of the whole vehicle, such as, the tightening accuracy deviation from $\pm$ 20% is reduced to within $\pm$ 5%. In addition, the AGV conveying technology realizes the self-adaptive change of the production line. Product changeover cycle is reduced from 15 to 5 days. The loss of production downtime and the investment in renovation are reduced.

In terms of information system, the factory is equipped with a newly developed MES system. The information in real time such as vehicle sequence models, process parameters, product quality, process data, energy consumption can be collected and analyzed to realize the full information closed-loop management. The information system is also equipped with a mobile APP, which facilitates daily communication management and improves the speed of information transmission and work efficiency.

In terms of defect management, the MES system is equipped with multiple subsystems such as torque management system, logistics management system, and Anton system, which interconnect information with each other, and cooperate with RFID technology and big data management to achieve traceability of problems and "zero outflow" of defects.

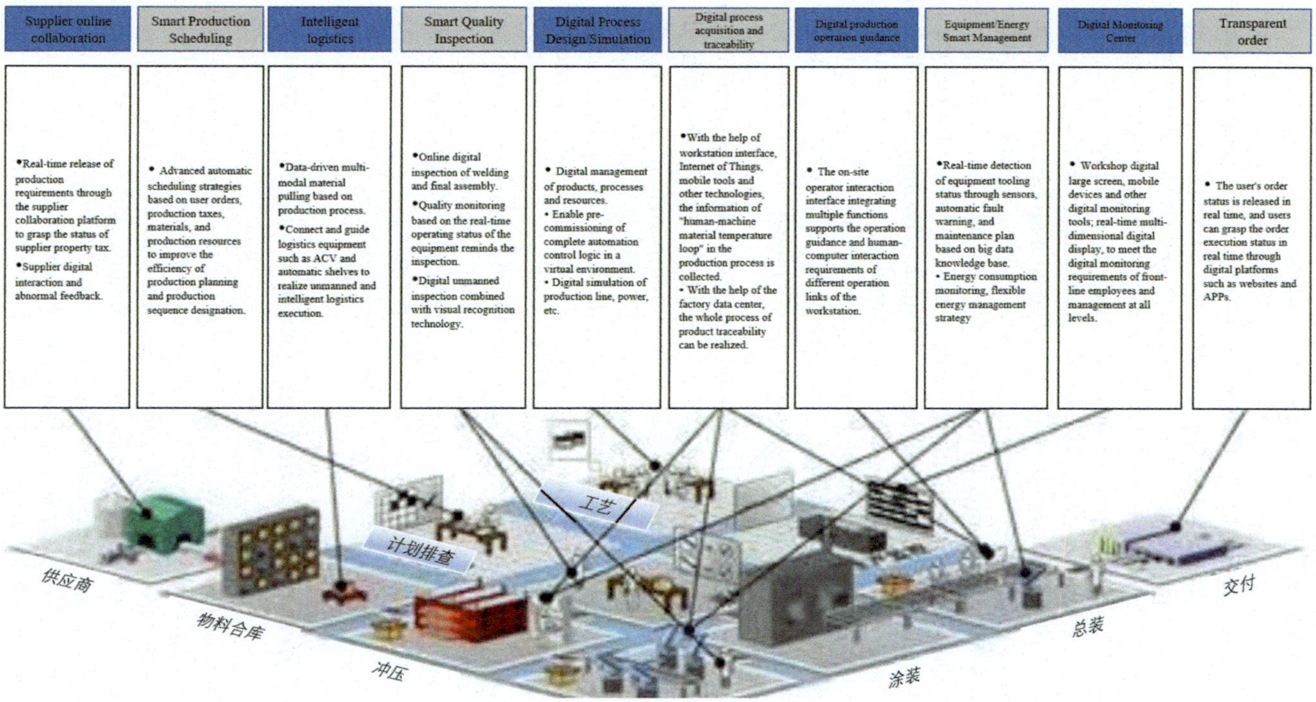

Fig. 6.26 Key scenes of China FAW digital factory

In terms of personalized customization, the expansion interfaces with sales order system, process design system are reserved to realize customized production to meet customer's arbitrary selection requirements, which greatly shortens the delivery cycle, and improves user experience.

In terms of logistics, the digital management of the entire supply chain is realized using the LES logistics execution system. The LES system consists of three modules, Kanban management, order segmentation and transportation monitoring. For planning, sending orders, in-transit transportation, in-plant storage, and on-demand online, the system automatically calculates the cycle and arrival time and writes them into the Kanban to achieve to true combining logistics and information flow, and achieve 100% "traceability of parts". The Hongqi LES logistics execution system innovatively realizes the intelligent segmentation of orders through the automatic calculation and distribution of the daily order volume. The required production batches can be adjusted at any time according to the actual situation of the factory and production line to reduce the number of in-stock. The implementation of lean logistics management methods such as "stack number" and "batch number" can accurately guide each operation in the logistics process, reduce handling and waiting links, and improve logistics efficiency. Parts are monitored in transit. The information of on-time rate and error rate are automatically uploaded to the system to realize digital management of parts.

In terms of internal logistics and distribution, the assembly workshop of Hongqi's new H platform adopts an unmanned and shed-free full-line SPS automatic distribution mode, and the online transportation of all 33 transportation routes is completed through nearly 160 AGVs in the factory. The LES system is used to read the lead time of material distribution to guide the logistics distribution rhythm, and the AGV will deliver the parts to the line side on time according to the system instructions. The introduction of RFID automatic identification technology can automatically identify the distribution locations of 38 kinds of materials and appliances in the field, and truly achieve intelligent logistics. It is the first general assembly workshop in China to realize unmanned distribution, which can effectively avoid wrong and missed installation and reduce logistics and distribution personnel.

6.5.2 The Effect of Implementing Smart Manufacturing

Since the reform of China FAW, a major change in the manufacturing field is to complete the transformation of traditional production methods and build a digital factory. The scope of construction revolves around 6 aspects of technology, planning, production, logistics, procurement, and quality, and 20 core processes in the large manufacturing field; opening up two core businesses of product development and order delivery; covering site, control, operation, factory management, enterprise management, Ecological collaboration has 6 system levels. It aims to create a digital factory with efficient and intelligent features, excellent quality, flexible customization, transparency and visibility, green environmental protection, and human–machine collaboration, and manufacture user-defined and user-satisfied high-quality products.

Through the implementation of intelligent manufacturing construction, the complete vehicle production and approval cycle has been shortened by 7 months, the order delivery cycle has been shortened by 26%, the frequency of claims per thousand vehicles has been reduced by 15.8%, the vehicle product review level has been raised from 1.5 to 1.4, and the single vehicle manufacturing cost has been reduced by 10%, the material inventory turnover rate increased by 50%.

GPSR Compliance

The European Union's (EU) General Product Safety Regulation (GPSR) is a set of rules that requires consumer products to be safe and our obligations to ensure this.

If you have any concerns about our products, you can contact us on ProductSafety@springernature.com

In case Publisher is established outside the EU, the EU authorized representative is:

Springer Nature Customer Service Center GmbH
Europaplatz 3
69115 Heidelberg, Germany

The manufacturer's authorised representative in the EU is Springer Nature Customer Service Centre GmbH, Europaplatz 3, 69115 Heidelberg, Germany. If you have any concerns regarding our products, please contact ProductSafety@springernature.com

Printed and bound by CPI Group (UK) Ltd, Croydon, CR0 4YY
01/10/2026
02225867-0005